DEFENDING LONDON'S RIVER

6 inch gun on the roof of Garrison Point Fort during the Second World War   *(Imperial War Museum)*

VICTOR T. C. SMITH

NORTH KENT BOOKS

FORWARD

by

Colonel B.E. Arnold TD RA (TA) Retd.
(former Commanding Officer of 410 Coast Regt. RA (TA)
and Chairman Kent Defence Research Group)

Not a day passes but some part of our history is uncovered. This is often brought about by enthusiastic young men and women who give their time voluntarily, exploring and researching into our past so that our heritage may be preserved and those who follow us in the years to come may learn how each generation lived, and its culture, especially in a world where one nation's hand was constantly raised against another.

The author is one of this breed of young men who devote much of their time in uncovering details of our past. Having armed himself with a considerable store of technical knowledge, he has applied himself to research and recording with enthusiasm, that part of our history relating to defence.

His task is endless for as soon as one fort, block house, or defence work is revealed, so another calls for attention. He has spent much time restoring the New Tavern Fort Gravesend built in 1780 and modernised over the years. It is described in detail in his book. The public may now enjoy strolling round this fort and breathe in the atmosphere which surrounds it and haunts its passages and corridors. His energies are now being directed to the restoration of Coalhouse Fort.

It is quite obvious to a reader that a very considerable amount of research, time and hard work has gone into the production of such a book and many hours spent in exploring the areas in the field.

Whilst the history of our Island race has been written on numerous occasions, this book, for the first time, provides a very detailed study of the defences of the Thames area and will be invaluable to students in the years to come.

Whether the reader be a student of history or just a casual reader, the book will prove fascinating and absorbing and I commend it as a welcome addition to any bookshelf.

Westbrook, Nr Margate, Kent — 1985

Cover picture
Bofors gun on one of the offshore forts (Army) during the Second World War (Imperial War Museum)

Published by North Kent Books, 162 Borstal Road, Rochester, Kent.
ISBN 0 948305 00 2
Printed by Albion Printers, Turkey Mill, Maidstone, Kent
© V.T.C. Smith 1985

Introduction

The artillery defences of the Thames, the most important river in the kingdom, have never been the subject of an account directed at the general public. Apart from the guidebooks to Tilbury and New Tavern Forts, the published matter has been restricted to the limited circulation journals of several archaeological and historical societies. This book is an attempt to fill that gap. It is a history rather than a guidebook and traces the development of these defences from 1540 until 1945. Included is an appendix giving brief historical summaries of the extant works, noting in general terms what can still be seen.

That the Thames led directly to the capital in which the political and economic life of the country was centralised was reason enough for providing it with defences against a hostile foreign power. More than this, in the event of invasion, an undefended river represented a tempting way into the interior of England which avoided an extensive overland trek from a coastal disembarkation. Added to this was the need to protect the very large amount of mercantile shipping moored in the river. Even in 1540, the security of the Thames involved vital interests: about 80 per cent of English exports passed out of the river and then there were the important royal dockyards at Deptford and at Woolwich to defend. The arsenal at Woolwich soon became the central store for England. Throughout the period covered, the defence of the Thames in time of war against a raid or a landing, was seen as being inexorably bound up with the security of the nation.

The first line of defence has traditionally been the navy but in case it was evaded or defeated there also needed to be gun defences on the shores of the river to bar entry. It was therefore logical that these shore defences should have been located as near to the entrance of the river as was possible. However, until the increased ranges and accuracy of artillery in the later nineteenth century, it was not feasible to attempt to fire on the navigable channel from batteries situated in the rapidly widening river below the Lower Hope — which was in any case provided with effective natural defences against a landing in the form of mudflats and sandbars, particularly on the south side. Therefore, for most of their existence, these defences were concentrated in the stretch of river where it first begins to narrow above the Lower Hope, notably at the eastern end of Gravesend Reach where the river bends, and at the western end of the Reach, before it bends into Northfleet Hope. Most of the developments in English artillery fortification have manifested themselves in the Thames defences in their long history and it is fortunate that much still remains to be seen today.

In the preparation of this book the writer has drawn upon extensive documentary research, site surveys and excavations which he has undertaken over the past fifteen years. However, space limitations have prevented the inclusion of some matter — such as the batteries which existed for a short time at Woolwich — and have hindered the exploration of some interesting themes. It is hoped that these, together with a more detailed examination of the defences during the two World Wars, can be covered in a future publication.

The writer would like to express a debt of gratitude to A.D. Saunders, MA, FSA, and J.D. Wilson, whose pioneer articles on the Thames defences which appeared in 1960 and 1963 respectively, provided the stimulus for the writer's further studies which have led to this book.

Northfleet — 1985   Victor T.C. Smith

The Precursors

Defences and defensive schemes of one form or another have been a feature of the Thames estuary since the Iron Age. The Roman occupation saw the construction of a fort at Regulbium (Reculver) which may have served as a base for naval patrols of the estuary while other fortifications may have existed at Hadleigh in Essex. Information about the medieval period, surely a fruitful area for research, is at present rather sketchy. However, from time to time, named officers responsible for defences appear in the records, such as Richard de Tany in 1295, John, Earl of Warenne in 1299 and William de Wauton in 1316, although defensive measures during this period could be as much against piracy as the organised attacks of a Continental state. Increasingly, in the fourteenth century, French raiding became a problem and an early-warning system of beacons was established to communicate news of an attack. One of these beacons was established at Roundonhyll (Windmill Hill) at Gravesend, and others at Tilbury and Shoebury.

Towers of Cooling Castle, showing key-hole shaped gun-ports (V.T.C. Smith)

The sheriffs of Kent and Essex were ordered to renew the beacon system in 1377 but whatever defensive measures were in force, they were to little avail, for in 1380 the possibility of an attack became actuality: in that year a combined French and Spanish fleet entered the Thames and landed at Tilbury and Gravesend, burning, looting and taking prisoners. It was the threat of such raids which helped lead to the construction of two castles — well away from the river itself, where the marshland joined the higher ground: at Hadleigh in Essex in the 1360s, and at Cooling in Kent (with its key-hole shaped gun ports, adaptations of the older arrow slit) during the 1380s. In 1402 the remote hamlet of East Tilbury vulnerably situated close to the river, received royal approval to erect a rampart and towers for protection against sea raiders.

Henry VIII and new fortifications, 1539/40

Castles were the badges of feudalism. Gunpowder did not of itself spell the end of castles, only the form of architecture which they employed. Rather the decline of the castle should be seen as running parallel with the decline of feudalism. Despite the turmoil of much of the fifteenth century, by its end English government was showing increasing tendencies towards centralisation, in tandem with similar trends in Continental Europe. However, it was not until the reign of Henry VIII that these tendencies were seriously translated into action in England's national defence. This followed growing concern in the 1530s about the inadequacy of the existing defences against the artillery-armed warships being built in increasing numbers by the Continental states. Added to this were the political dangers resulting from Henry's quarrel with the Pope. Concern was brought to a head in 1538. In that year the rivalry between France and the Habsburg Empire for the domination of Europe which Henry had been able to manipulate for England's safety by backing one side against the other, was replaced by a Franco-Habsburg truce. This left England politically isolated and the Pope seized the opportunity to try to organise an invasion to restore the country to the authority of the Roman See. The English response to this threat was to expand her fleet and to begin a major programme for the construction of forts and batteries at strategic points on the vulnerable 'invasion coastline' from Hull to Milford Haven.

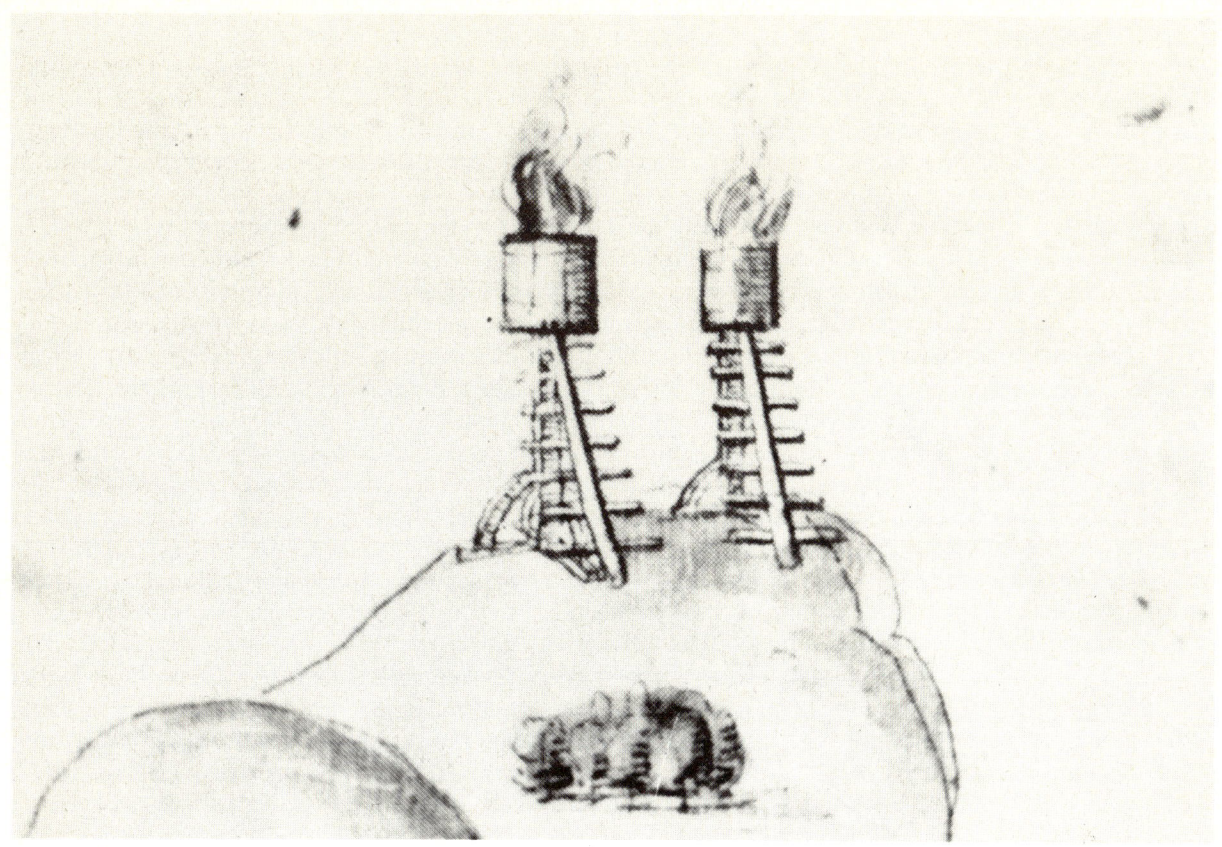

Beacons, possibly like the ones on the banks of the Thames   *(Part of B.L. Cotton Mss Aug I/i, 33)*

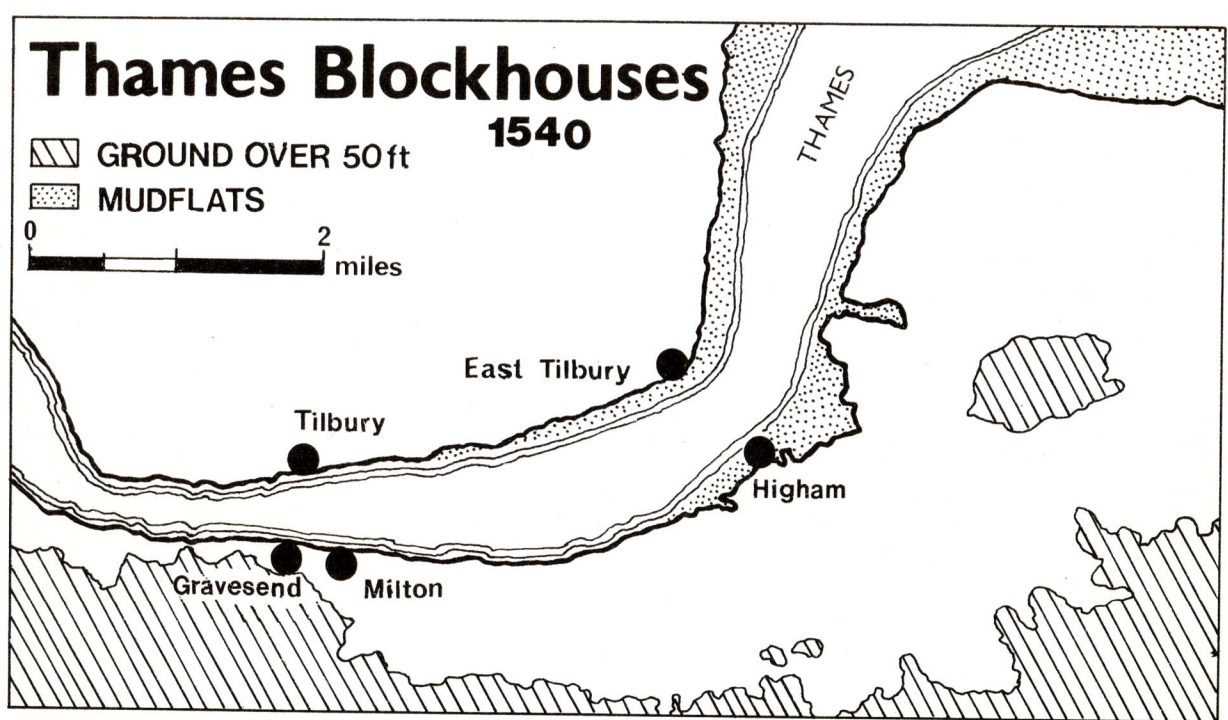

The Thames defences were an integral part of this defence programme. Where previous development in England had, with few exceptions, responded to gunpowder and artillery by mere adaptation of medieval architectural tradition, the fortifications built under this scheme were fully-fledged and purpose-built artillery forts designed to fight and survive a stand-off artillery duel with a fleet. Unlike the castle which combined a lordly residence with a military function, the forts were exclusively military and royal. They were low in profile, with the guns firing from specially designed emplacements in bastions having a curved front to deflect shot. Where previously it had been usual to purchase artillery from abroad, Henry placed great importance on England starting to develop a self-sufficiency in ordnance manufacture. He also encouraged the study of gunnery by native-born Englishmen.

As part of the national plan, in 1539/40 five artillery blockhouses were built near the mouth of the river Thames. These were to the design of James Nedeham and Christopher Morice, both of whom had previously served with the King's armies in France and had been involved in the construction of fortifications there. They were sited where the river first begins to narrow after the estuary: two crossed their fire at Higham (Kent) and at East Tilbury (Essex) and formed an outer line of defence; the three others at Gravesend and Milton (Kent) and at Tilbury (Essex) formed an inner line some 5,000 yards upstream and guarded the all important strategic Gravesend/Tilbury ferry crossing. The plans of the Gravesend and Tilbury blockhouses, the only ones known, show them to have been D-shaped structures of brick, two storeys high, with guns mounted in bomb-proof casemates in the semi-circular front and in the open on the roof. There were other positions for guns on adjacent ramparts on the riverbank. The guns — between 25 or 30 at each blockhouse — were a mixture of types; some were breech-loading and others muzzle-loading; some were of iron and others of brass; calibres varied a good deal from a mighty 9-in. bombard to a tiny 1½-in. Falconet. Even at this period recent improvements in manufacturing techniques and in powder quality meant that some of the guns had an extreme range of well over a mile, although effective ranges were much shorter. Despite the relatively inaccurate guns then in use, the shore battery had a considerable advantage over the unstable gun platform offered by a ship bobbing up and down on the water. The stability of the land gun platform gave the defence more chance of obtaining an accurate hit whilst a ship could not achieve a superiority of fire until it was broadside on and could bring all its guns to bear. By then, the attacking vessel or vessels may have suffered considerably from the land-based firepower. Close defence against landing parties was provided for by a small issue of handguns, pikes, bills and even bows and arrows. During the excavation of the Gravesend blockhouse some cross-bow bolts were found.

Permanent garrisons were quite small — a captain, his deputy, a porter and half a dozen or so gunners and soldiers. They were paid out of funds provided by the Exchequer and the Office of Ordnance. At one shilling per day, the captain received only twice as much as the lowest paid in his command. It is evident that later in the century some of the gunners, such as Jasper May and

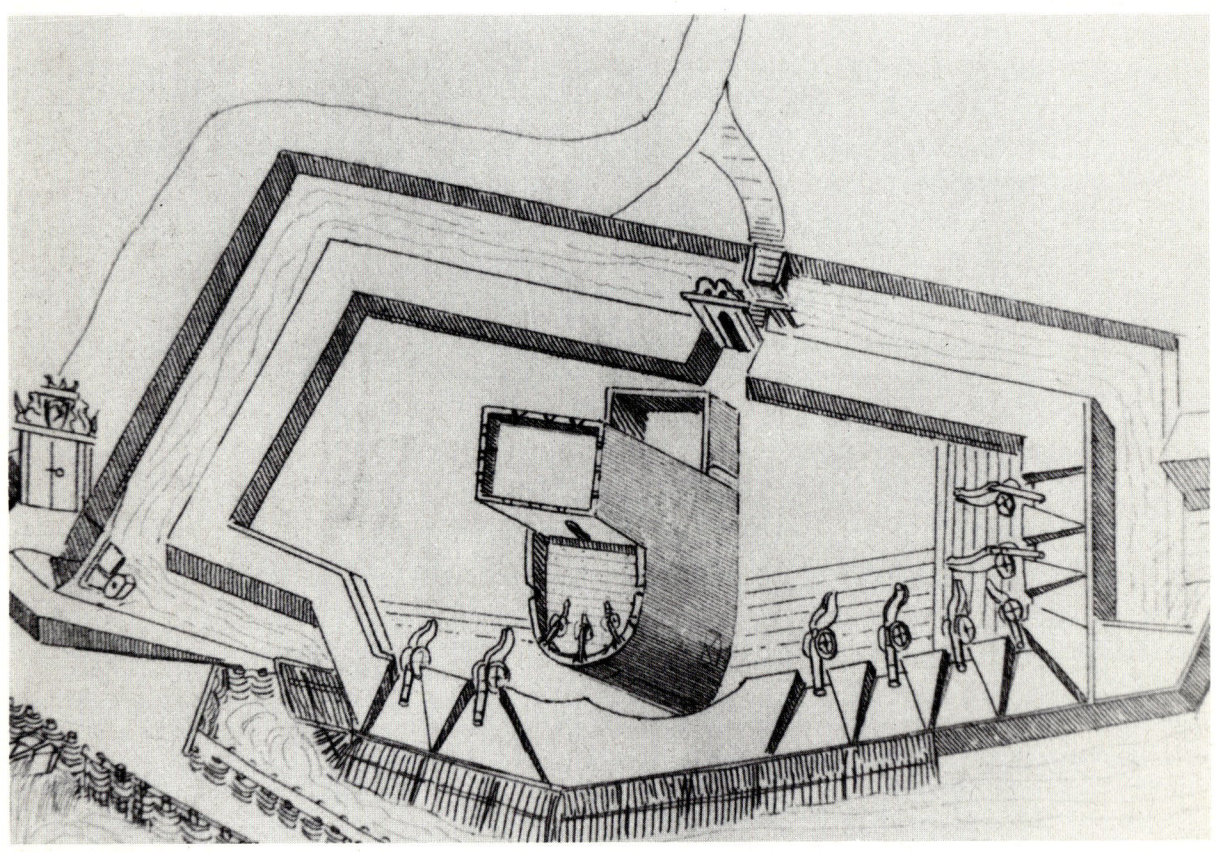

Tilbury blockhouse in 1588 but showing it as it probably appeared in 1540
(Cruden's History of Gravesend, 1843)

William Bourne (also a mathematician) at Gravesend, eked out their living by sharing their military duties with a part-time occupation such as boat building and inn keeping. The garrisons would have been enlarged in periods of emergency.

The foundations of the Gravesend blockhouse, excavated in 1977 and restored at the prompting of the Kent Defence Research Group in 1980, may be seen in front of the Clarendon Royal Hotel. Other excavations undertaken by the Kent Defence Research Group on the site of the Milton blockhouse in 1978 failed to reveal the original blockhouse but did find a rear extension in the form of an angular bastion of which only the chalk foundations still remained. Records evidence suggests that the bastion had been designed in 1545 by Richard Lee, a well-known military engineer, who designed the similar but later angular bastion at Upnor Castle (1558). The Milton structure must be counted among the earliest angular bastions built in England. Its remains were back filled so that nothing is now visible, although a plaque marking its site is proposed.

Following disarmament in 1553, Milton and Higham blockhouses were demolished in 1558 to provide materials for repairs to the Tower of London. Documents of the period show that a reward was given by the Crown for information about the theft of lead from the roofs of these blockhouses. East Tilbury blockhouse soon faded into decay and obscurity although some of its site appears to be extant and would merit archaeological investigation. By the eighteenth century the position of Higham blockhouse had been scoured by the river and only Gravesend and Tilbury blockhouses remained in commission.

The Spanish Armada, 1588

For the early part of her reign, Elizabeth I played a skilful diplomatic game with France and Spain which succeeded in keeping England out of direct involvement in Continental wars. As a result, these were quiet years for British defences and the Gravesend and Tilbury blockhouses which were allowed to slip gradually into decay. When Elizabeth's action in assisting the disaffected Netherlands helped provoke Spain into launching an invasion against England in 1588, the Thames blockhouses were in a hopelessly unprepared condition. The designated local commander, the Earl

Semi-circular front of the Gravesend blockhouse (V.T.C. Smith)

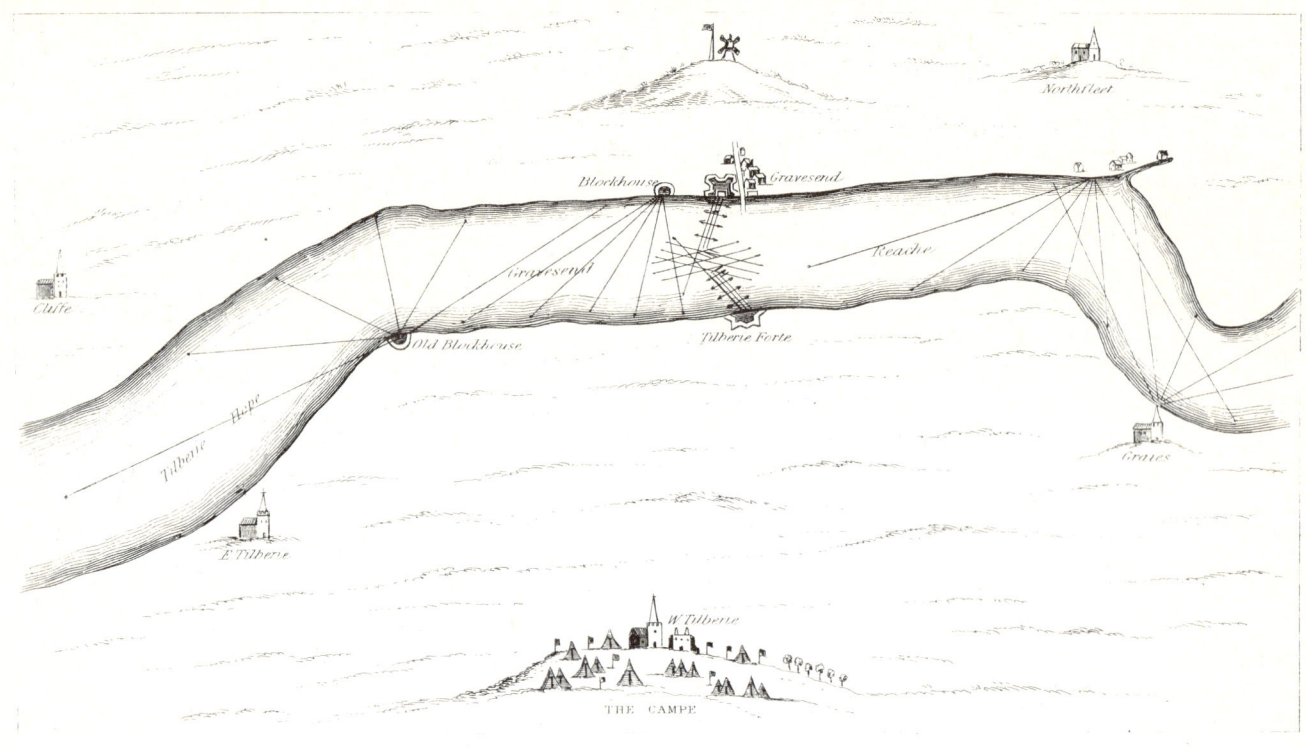

Defence of the Thames at Gravesend in 1588   *(Cruden's History of Gravesend, 1843)*

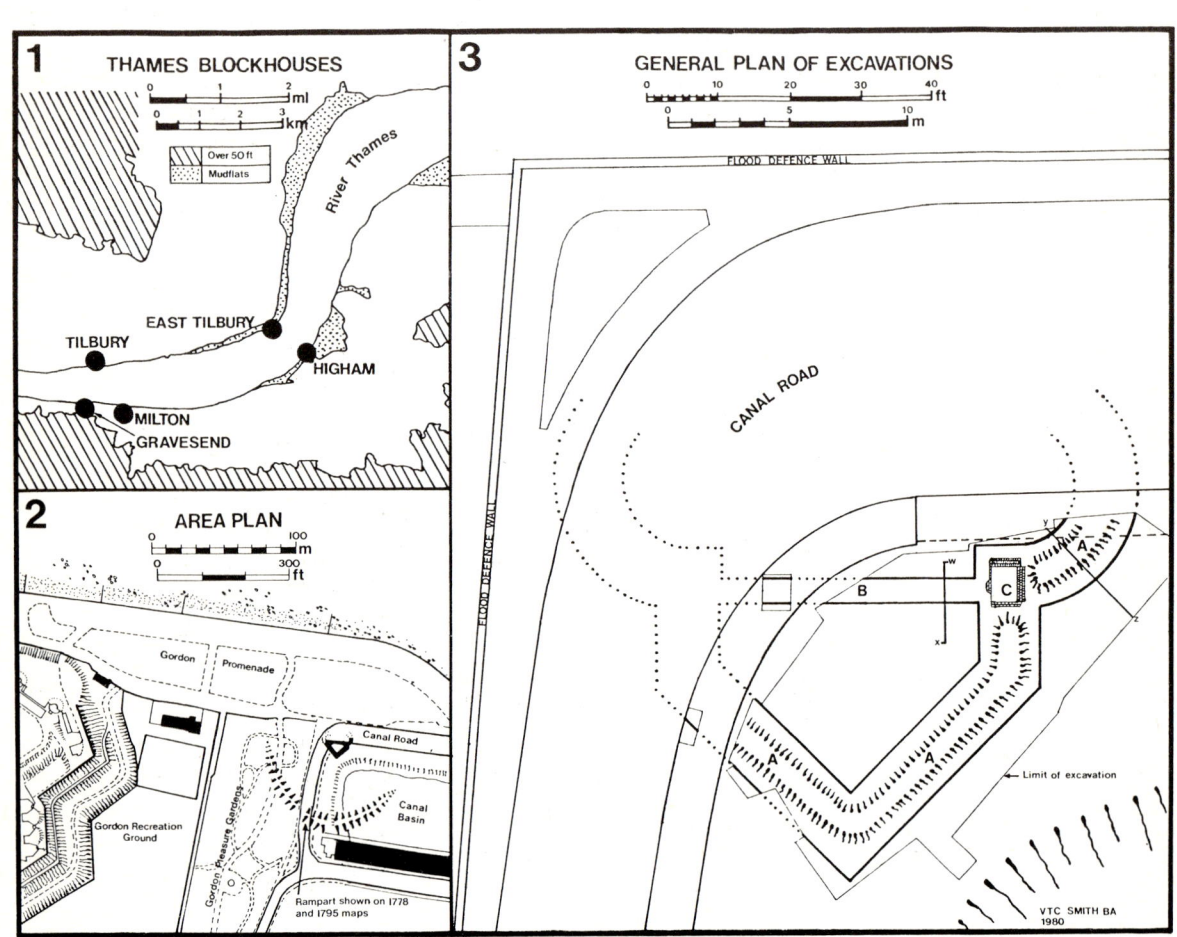

Milton blockhouse: Location and Plan

8

of Leicester visited them in July, and found them without 'one platform, either on the ground or aloft to bear ordnance', and with few guns and most of those on rotten carriages. Since at that moment the Duke of Parma had massed a force of 25,000 men in the Netherlands for an assault on the Thames, this was an extremely dangerous situation which called for swift and vigorous remedial action.

Work on repairs proceeded as quickly as time would permit — at Tilbury the blockhouse was apparently enclosed within a new star-shaped rampart and ditch on the landward side and there is some evidence to suggest that an outer rampart was added on the high ground immediately to the rear of the Gravesend blockhouse, in an area later known as the Camps. Passage to shipping was not only to be hindered by gun-fire but physically blocked by an 800-yard long boom which was strung out across the river between the blockhouses. Part of the work at Tilbury was organised by the imported Italian engineer, Federigo Gianibelli. He was employed by Elizabeth on a number of fortification projects but he was perhaps best known at the time for his 'infernal machines' or giant floating mines with which he had blown up the Spanish boom defence and its 800 defenders at Antwerp in 1585. His own attempt at a boom defence between the Gravesend and Tilbury blockhouses in 1588 was not successful as it broke up on the first flood tide and was replaced by one devised by an Englishman, Peter Pett, from the Royal Dockyard at Deptford. Surprisingly, there seems to be no clear evidence of any steps taken to re-occupy forward positions such as at East Tilbury and Higham although East Tilbury blockhouse is certainly marked on a map of 1588. Against the possibility of a landing, a field army of 23,000 men — one of four armies deployed in the South East — was encamped on a plateau overlooking the river at West Tilbury, available to cross the river into Kent as required. It was here on 8th August, that Elizabeth gave her famous morale-building speech to the assembled troops. As well as the lower Thames, the upper river was protected with batteries.

Fortunately the Spanish forces had failed to secure for themselves a deep-water port in the narrow seas from which they could embark Parma's army and the defeat and dispersal of the Armada at sea finally dispelled any immediate fears of invasion. Although the camp at Tilbury dispersed, the unfinished repair work to the blockhouses continued for several months, and the accounts refer to an increase in wages to the labourers 'by reason of the watry colde and fowlenes of the worke'. Unfortunately, no traces of the Armada preparations are now discernable.

It is clear that contemporaries drew no long-term comfort from the defeat of the Armada and saw it merely as a temporary breathing space while Spain rebuilt her fleet. Various anti-invasion contingency plans were drawn up for deploying warships and troops for the defence of the river, for a revived early-warning system of signal beacons and message carrying barges. Even as late as 1600 there was a Spanish invasion scare. However, after the war with Spain came to an end in 1604 the Spanish invasion threat ceased to exist.

Entrance to Duke of York's quarters, now incorporated into the Clarendon Royal Hotel
(Cruden's History of Gravesend, 1843)

The early Stuarts and the English Civil War

The reigns of the early Stuarts were, in general for English defences, a dismal picture of neglect, garrisons remaining unpaid for years and works falling into decay, as evidenced in the Thames by the several surveys of the blockhouses in the period which recommended extensive repairs. During the English Civil War, the Thames blockhouses had a rather unglamorous and undramatic part to play: at the outset, Parliament had seized control of most of the forts and garrisons in the South East and the blockhouses merely served out their time as check-points for shipping entering and leaving the river, the commanders having special powers to detain persons suspected of being counter-revolutionaries or Royalist agents. London itself was fortified during the Civil War by an 11-miles long defence line but this was chiefly against land attack by Royalist forces. On the Restoration, extensive accommodation was added in the form of a detached building to the rear of the Gravesend blockhouse for the use of the Duke of York (later James II) as Lord Admiral. The front of the building was appropriately decorated with an anchor and semisphere above it, with the date 1665, all in brickwork over the door. This building, which later became the ordnance store keepers' quarters, still exists in an extended and altered state as the Clarendon Royal Hotel.

The Dutch Raid, 1667

Unfortunately, the provision of comfortable and suitable accommodation for the Duke of York was not matched by expenditure on maintaining the blockhouse itself. When the Dutch under Admiral de Ruyter raided the Thames and Medway in 1667 during the Second Dutch War the blockhouses were in just as poor a state as they had been on the eve of the Spanish Armada. The event is chronicled in some detail in the state papers and in the Diary of Samuel Pepys: the gun platforms were in an advanced state of decay and all that stood between the Dutch and a clear passage upriver were half a dozen small and rusty guns on rotten carriages. That the Dutch never actually tried to force their way through was the result of a misapprehension — they believed the blockhouses to be well armed and did not wish to be caught in a cross-fire. This bought time for other guns — some 80 of them — to be brought in from elsewhere, for repairs to works to be effected, extemporised works to be constructed and for a flotilla of warships to be moored in line between the two blockhouses. Batteries were also built upstream at Woolwich. Although these measures blocked an advance upriver, while the Dutch were causing havoc in the Medway, de Ruyter did launch a marauding attack on Canvey Island where a few barns were burnt. East Tilbury Church is said to have been damaged by gun fire during the Dutch Raid. Fortunately, the Peace of Breda in June, 1667, brought the war with the Dutch to an end.

Early nineteenth century view of Gravesend blockhouse and storekeepers quarters (ex Duke of York's) behind
(Gravesham Borough Library)

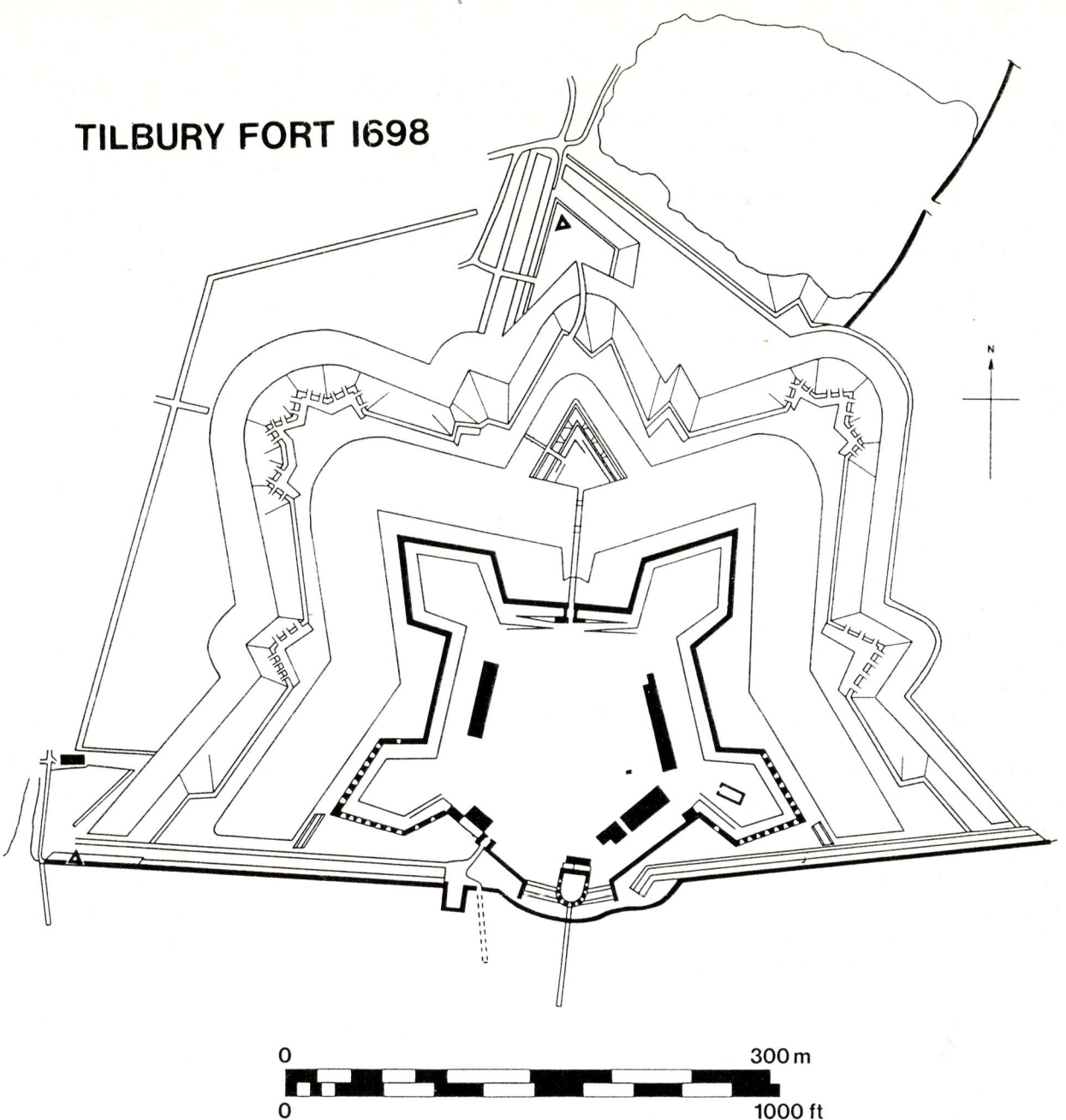

(P.M. Wilkinson, Excavations at Tilbury Fort, Post Medieval Archaeology, 17 (1983))

The bastion trace and the building of Tilbury Fort

The inadequacy of the blockhouses during the Dutch raid had focussed attention on the need to improve the Thames defences and this led to the building of Tilbury Fort around the old Tilbury blockhouse. However, this should not be viewed simply as just a case of cause and effect: plans had been made earlier for a new fort — in 1661 and 1665. Rather Tilbury Fort should also be seen as part of a wider strategy to improve the nation's defences, which included new works in the Medway, at Plymouth, Portsmouth and at Harwich.

The fort, an extensive pentagonal bastioned work, now preserved in the care of English Heritage, must be counted as the finest remaining example of the work of the seventeenth century military engineer in England. It was built between 1670-84 to the design of a Dutch engineer, Sir Bernard de Gomme. Whilst the angular bastion had already manifested itself briefly in the Thames in the single bastion grafted on to the rear of Milton blockhouse in the 1540s, the new Tilbury Fort was a complete defence system in which the angular bastion was integral.

This form of fortification, known as the bastion trace, originated in late fifteenth century Italy. Perhaps the most significant new architectural development of the Italian Renaissance, it was a development which followed on from the earlier rounded bastion system which was less efficient and produced dangerous areas of dead ground. It was not employed in England until near the end of

Tilbury Fort – one bastion commanded the ground in front of another   *(V.T.C. Smith)*

Water gate at Tilbury Fort   *(V.T.C. Smith)*

Henry VIII's reign, after the main phase of his fortification programme was complete, but was extensively used thereafter. Basically, the area to be defended was surrounded by a rampart on a polygonal plan and an angular bastion was placed at each point of the polygon. This configuration practically eliminated areas of dead ground and therefore ensured, for close-defence purposes, that no point in the vicinity of these ramparts was uncommanded by fire whilst retaining provision for long range fire. Often this system was applied concentrically, with intervening ditches and works, giving successive lines of defence – hence 'defence in depth'. By the seventeenth century the bastion trace had reached considerable sophistication, famous engineers like Coehoorn in the Netherlands and Vauban on the frontiers of France, cunningly contriving it into what seems a geometrical nightmare, with all manner of outworks and retrenchments. Even in the seventeenth century, England was still very much a receptacle for Continental ideas and whilst by no means devoid of experience in the art or science of fortification, tended to favour employing Continentals to design and to supervise the more important works, such as Tilbury Fort. The choice of de Gomme, who had served the king during the English Civil War, was therefore a natural one and it was equally understandable that Sir Martin Beckman, another Dutchman, should succeed him as king's engineer when he died.

The conditions at Tilbury – a low, level, marshy site – were almost 'Dutch' and de Gomme's Continental experience and that gained in the service of Charles I at Oxford during the Civil War, was put to good effect in his planning of the defences. Tilbury Fort was constructed of earth with massive brick revetments and was protected on its landward sides by two wet ditches. The inner one contained a triangular island called a ravelin to block fire against the north curtain and the outer one a tiny triangular brick redoubt (later enclosed in an earthen redan). The counterscarp of the inner ditch was provided with a covered way and gun positions. The ditches could be drained in very cold weather to avoid ice forming which could be simply walked over by an attacker as had happened earlier in the case of some fortifications in the Netherlands. The land access to the fort was formed by bridging over the ditches but 'draw' sections were provided so that the garrison could cut the way. The bridges, which took an indirect route from the redan, via the ravelin to the Landport Gate, have been replicated by English Heritage and give the visitor, on entering the monument, a good immediate appreciation of seventeenth century 'defence in depth'.

As to the riverfront, where the old blockhouse was retained, a fifth bastion intended to project out into the water was started but not finished. However, its shape can be worked out from the piles which still exist in the river. Gun lines along the river bank in front of the fort were formed to rake ships trying to get past. The riverside entrance to the fort was through the Water Gate, still existing, which has a very fine and ornate edifice of Portland Stone above it, formed of trophies of arms.

The interior of the fort contained barracks on either side of the parade and other buildings such as magazines.

Because of the marshy nature of the site, much of the structure had to be piled, a good deal of the timber for this having been imported from Norway. Many tons of bricks were left over at the end of construction and these were dispatched to Greenwich for the building of Flamsteed House, which became the Royal Observatory.

Cannon on field carriage at Tilbury Fort *(V.T.C. Smith)*

One of the drawbridges at Tilbury Fort *(V.T.C. Smith)*

Seventeenth century drawing of Gravesend blockhouse by Claude de Jongh   (B.L. Royal 6 ECM 6)

The developments at Tilbury overshadowed the modest Gravesend blockhouse which was little altered during the period. However, it was from this time that the first perspective drawing of the blockhouse exists. Some interest is also provided by the discharge of the governor at Gravesend, Sir John Griffith in 1669, for demanding money at gun point from ships before he would let them pass. The practice of exacting fees seems to have been adopted on various occasions by different governors, only to be stopped following complaints but resumed after the tumult had died down. The state papers record that in 1690 an Edward Lawrence claimed the credit for saving Tilbury Fort and the Gravesend blockhouse from an attack by 900 Irish. These were presumably soldiers but whether the cause of the apprehended attack lay in wider politico-religious grievances or in some other issue is not yet known.

Since the East Tilbury and Higham blockhouses had been abandoned in 1553, the Thames had been without any permanent provision for forward defence. This weakness revealed itself very clearly in the Dutch raid when the Dutch vessels were able to linger in the Lower Hope without hindrance or danger. Indeed one of the contemporary criticisms of Tilbury Fort was that it should have been built downstream at East Tilbury, to close the entrance to Gravesend Reach. Yet this obvious defect was to remain for a further period of more than a hundred years.

Although by no means free from invasion scares as conflicts with Continental powers occurred, the century following the building of Tilbury Fort was essentially one of adjustment and consolidation for the Thames defences. Following the Treaty of Utrecht in 1713, the armament of Tilbury Fort was reduced from 161 to 60 guns and the Gravesend blockhouse from 17 to 10 guns. The Gravesend blockhouse like its counterpart at Tilbury, was converted into a magazine. At Tilbury Fort a pair of very capacious magazines were also added on either side of the Landport Gate in 1716. In the previous year Tilbury had served briefly as a prison camp for prisoners taken in the Scottish rebellion. But for much of the period the fort served merely as a transit camp for various regiments and this is attested to in the burials of men from many units in local parish registers. Tilbury Fort was not a very pleasant place in which to live. Gravesend blockhouse had the advantage of being close to the amenities and diversions of the town but Tilbury was on a damp, marshy and malarial site in which even such a basic requirement as fresh water was not easily obtained. Certainly the officers at Tilbury preferred to live in Gravesend if they could. The available records give a fascinating insight into the routine of garrison life. Letter books record all the requisitions for supplies, repairs, and disputes about grazing rights over a considerable period, noting in 1732 that one stormy night the soldiers' boghouse at Gravesend 'fell into ye Thames'. Tilbury Fort had its own excitement in 1776 when a dispute at a county cricket match between Kent and Essex, played on the parade ground, led to bloodshed. Some of the players seized arms and bayonetted one of the soldiers trying to restrain them and shot dead a sergeant.

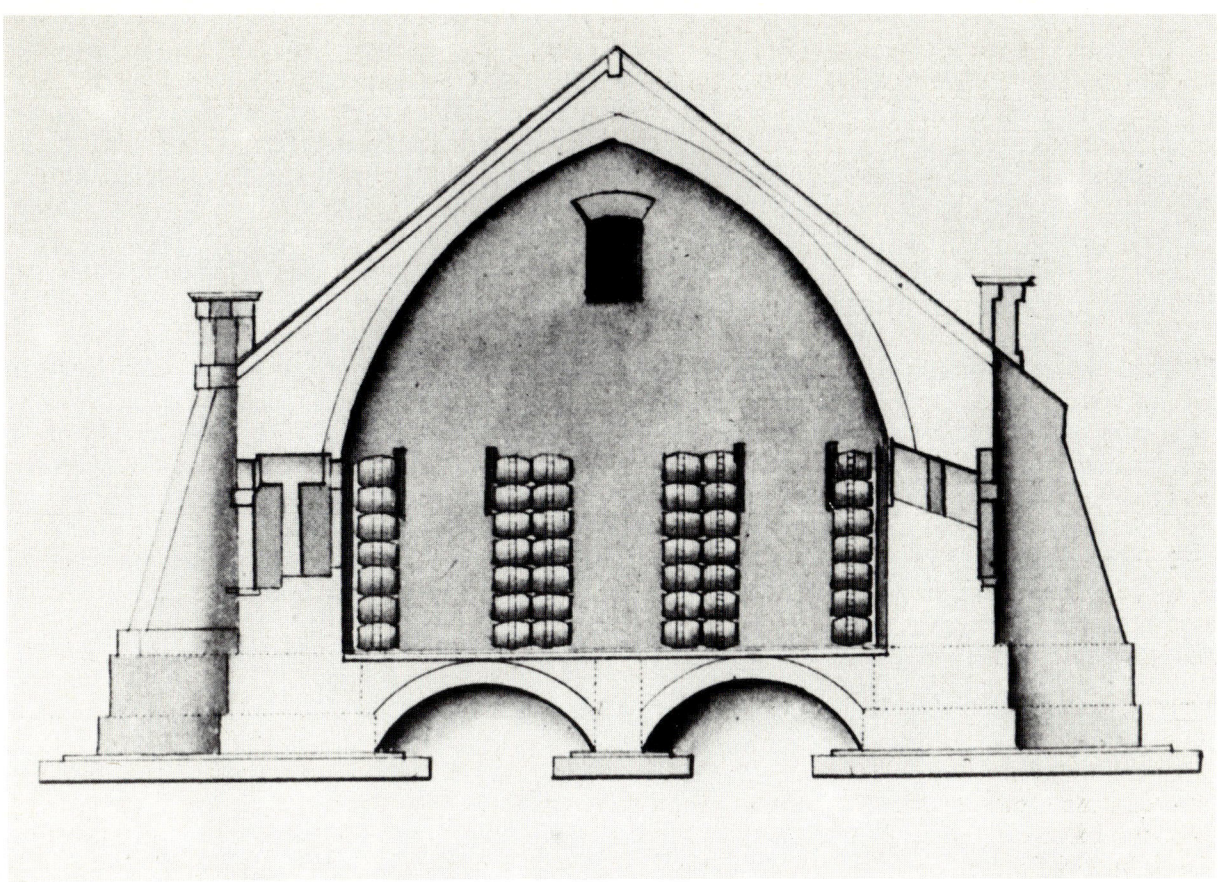

Section through one of the new magazines built at Tilbury Fort in 1716   *(Public Record Office)*

The building of New Tavern Fort, 1780

Wars with France and French invasion scares seem to have been one of the most recurrent themes of British history in the later-modern period. When France allied herself with the insurgent American colonies in 1778, the invasion scare which resulted gave rise to the first significant new building in the Thames for a hundred years — the construction of New Tavern Fort at Gravesend. However, despite the important target offered by the Purfleet magazine complex established only 20 years before, it was almost two years before work actually started. In the meantime, a new river communication was established between Gravesend and Tilbury for the movement of troops between the two counties in the event of invasion. The communication consisted of three jetties on either side of the river connected by hawsers which were used to warp six barges back and forth across the water as required. Each barge had a drop down end, rather like a modern landing craft. The communication was tested in 1780 when Lt. Gen. Pierson embarked a force of several thousand men for a mock attack on Tilbury Fort. All these facilities had to be judiciously guarded and the records note that on one stormy night a sentry did not leave his post at Gravesend until the rising tide had reached his knees while a sergeant was commended for rescuing a sentinel who had been washed downstream in his sentry box. The communication was connected on the Essex side to a military road from the hinterland. This was built by a combination of military artificers and civilian labour squads, like the 'Scum of St Giles' who terrorised the inhabitants of Brentwood in their off-duty hours.

New fortifications in the Thames area were merely one part of a much wider national programme undertaken at this time. In June 1778, all the English defences were surveyed and orders were issued for the construction of over 30 new batteries along the British coasts and for existing works to be strengthened. In the Thames, the main effort was directed at improving the defences on the south bank at Gravesend, although Tilbury Fort did receive some repairs, and a new six-gun battery was formed in the counterscarp of the inner ditch on the east side of the fort. The idea of a water bastion at Tilbury was briefly resurrected but it was not actually built. At Gravesend, however, the old gun lines immediately to the east of the blockhouse were completely remodelled to take an armament of 19 heavy cannon while just a little further to the east an entirely new battery was built for 16 guns. The latter was named as New Tavern Fort after an inn of that name on the site. The fort was constructed of unrevetted earth and was designed for an armament of heavy cannon firing through embrasures. Of an irregular plan, it consisted of a battery on two faces forming an angle towards the river, with a strip of rampart joining it to a smaller straight battery facing east. It was defended to the front by a flat-bottomed wet ditch containing a wooden palisade ten feet high. Unfortunately, the gun emplacements have been obliterated by subsequent reconstruction. Milton Chantry, part of the 'New Tavern', within the fort area was encased in brick and used as a barracks. Except for a few buildings, (e.g. a house transferred from an adjacent site on rollers and used as the Commanding Engineer's residence) the rear of the fort was open and consequently unprotected. However, before the end of the century the rear was closed with a brick wall loopholed for musketry defence. The cannon with which the fort was armed (in 1805 they were 2 x 3-pdrs., 14 x 24-pdrs. and 1 x 9-pr.) had an extreme range of well over a mile and most were probably mounted on Naval-style wooden truck carriages. There was also a shot kiln conveniently nearby. In this, round shot were heated until red hot when they could be fired at enemy warships to set their wooden hulls on fire. Red hot shot was found very effective during the British defence of Gibraltar in 1782 when they were used to set on fire the floating batteries of the besiegers.

The proposals of 1778 also included contingency plans for turning Gravesend into an entrenched camp in the event of invasion by forming extensive landward defences to the rear of the town, pivoting on Windmill Hill. Military camps already existed at Warley, at Dartford and elsewhere, ready to deploy forces to where they might be needed.

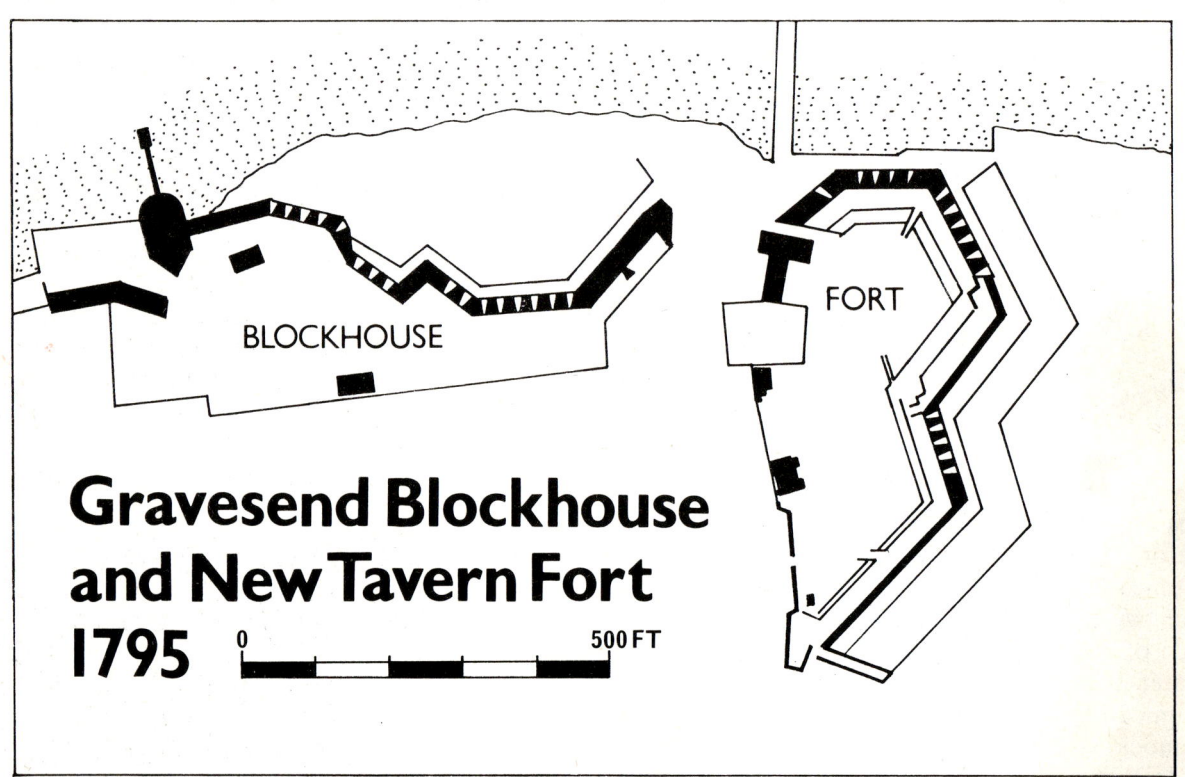

The French Revolutionary War and a gap plugged — the building of the forward batteries, 1796

Just as the building of Tilbury Fort in 1670-84 had done nothing to guard the vulnerable lower end of Gravesend Reach, so the limitations of the new works in 1780 to Gravesend and Tilbury demonstrated a persistent failure to appreciate the value of forward defence. It was only in 1794, during the French Revolutionary War, when the question of the defence of the Thames was comprehensively studied by Lt. Col. Hartcup, that action was taken. In the following year, a small expedition fitting out in Dunkirk was thought to be earmarked to invade England through Essex. In fact it was intended for the Tyne or Humber but never sailed. Nevertheless, there was an attempted invasion of Ireland that year and an actual landing took place at Fishguard in 1797. All this had emphasised the need for new defences. Hartcup's scheme called for three new batteries at Coalhouse Point in Essex and at Shornemead and Lower Hope Point in Kent, which were built over a

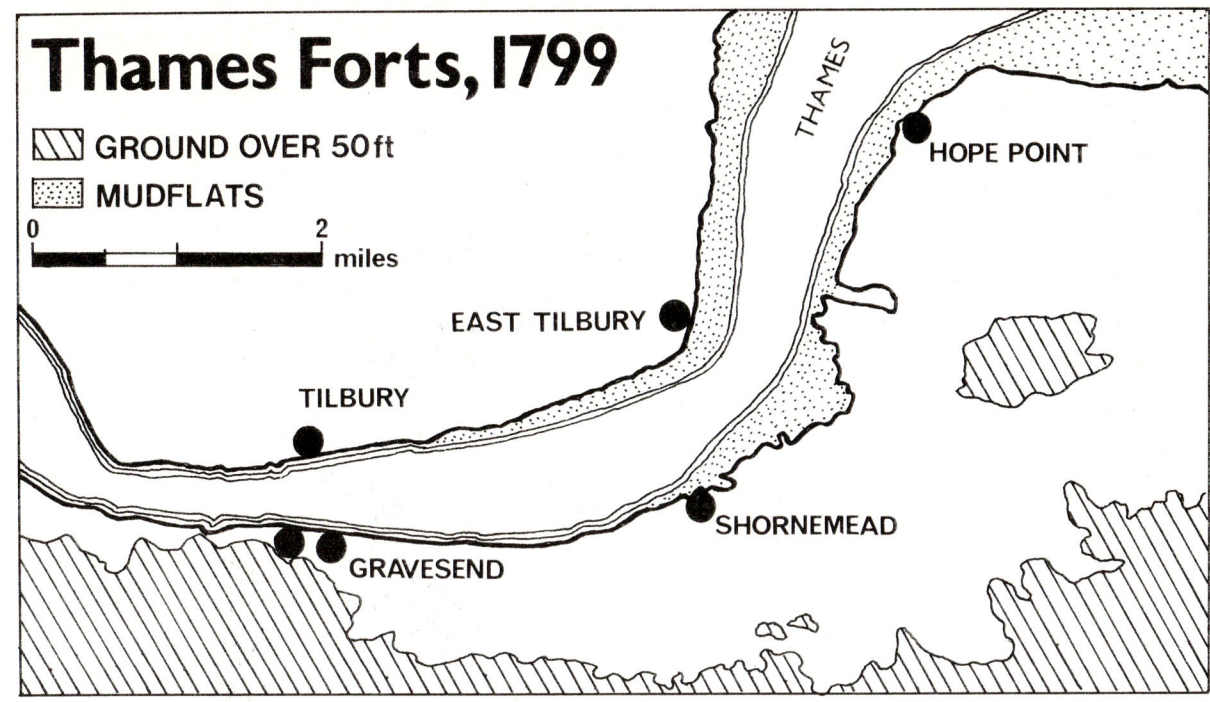

period of several years starting in 1796-9. They were of semi-circular plan, having a wide earthen rampart to the front and a triangular shaped walled off area to the rear, containing a weatherboarded barrack and a magazine. A fire-step was provided behind the enclosure wall whose faces could be flanked from musketry parapets either side of the base of the triangle. Outside the enclosure was a shot kiln. The irregularly shaped dyke which surrounded each battery was presumably mainly intended to drain the water-logged site. The latter produced its share of problems during construction of the batteries, causing slippages, shifting foundations and cracked structures.

Four 24-pdr. cannon were mounted at each work. However, they were not mounted on the old style naval pattern truck carriages but on a new type of traversing platform, which pivoted on rails and was more quickly and easily turned to pace a moving target. It might be supposed that a dozen guns were too few to defend the mouth of the river against what might prove to be a sizeable enemy force but on a number of occasions it had been shown that a few well placed guns on stable land positions could defeat a much larger number of guns mounted on an unsteady platform bobbing up and down on the water. At long ranges the land-based gun had decided advantages over the waterborne one and, as previously mentioned, only at relatively close quarters was shipboard armament likely to prove itself superior.

Warships moored across the Thames at the Hope, 1804.

17

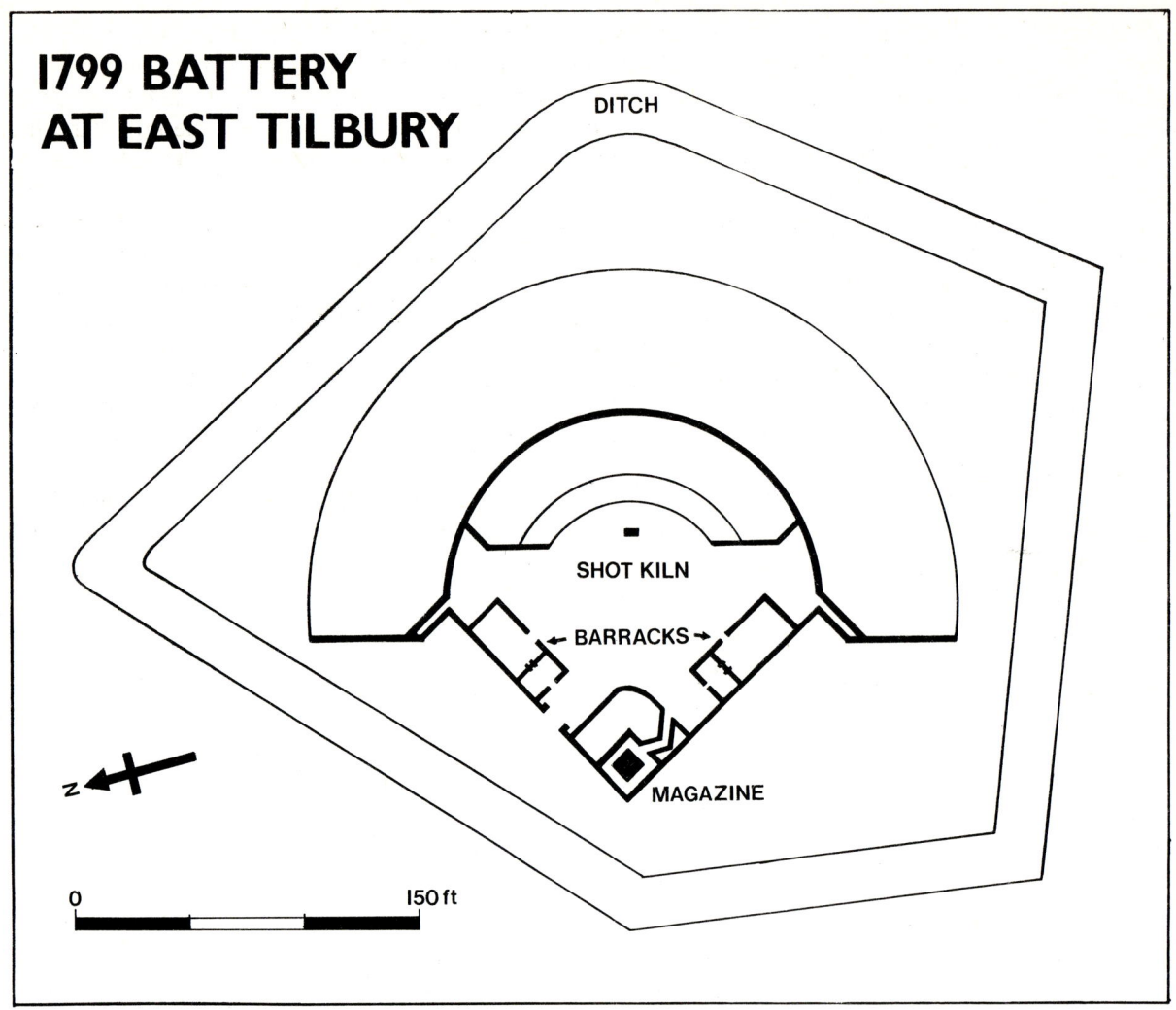

The semi-circular front and triangular rear seems to have been a favoured plan for new batteries at this time, several other examples being known in Kent at Dungeness and elsewhere.

There is evidence of a small battery having been constructed by a Major Birch upriver near the India Arms at Northfleet in about 1795. It was apparently to have received 2 x 9-pdr. cannon but does not seem to have lasted for very long.

There was also a supplementary defence in the form of a gunboat manned by volunteer seamen — later reinforced by a captured French frigate.

During the Nore Mutiny of 1797 there was some fear that the mutineers might enter the Thames and the forward batteries were put in a state of defence. The battery at Shornemead even fired a shot at a boat carrying a delegation of mutineers to the shore nearby.

During 1802, an attempt was made to drive a tunnel under the Thames between Gravesend and Tilbury. This was intended to form both a secure military communication for troops and a route for commercial traffic. Although the scheme was supported by the Board of Ordnance, it was privately funded and organised. Through mismanagement the scheme never got beyong the sinking of a deep pilot shaft at a riverside spot on the west side of Gravesend.

The Napoleonic Wars, in a sense a continuation of the Revolutionary Wars under new management, saw no new building in the Thames, although the gun positions of the forward batteries were raised to give greater command and some repairs were undertaken at New Tavern and Tilbury Forts. During an invasion scare of 1804 however, a flotilla of warships was moored across the Lower Hope in support of the batteries. Although no attack came, in 1806 New Tavern and Tilbury Forts were assailed by enemies of a different kind — a huge number of moles which attacked and threatened the structural integrity of the earthen ramparts. The mole catchers employed to counter the problem seem to have been successful but no sooner was their work completed than both forts were infested by swarms of black rats.

The defeat of the combined French and Spanish fleet at Trafalgar in 1805 did not in contemporary eyes end the threat of invasion but merely postponed it. The continued expansion of the French fleet down to 1814 lent support to British fears and the construction of such defences as the Martello

Towers and Royal Military Canal continued. Fortunately, throughout hostilities Britain and her allies were able to maintain naval supremacy which helped insulate the Thames against any possibility of attack. One of the last acts of the Thames forts was to salute Louis XVIII as he sailed out of the Thames in 1814 to resume the French throne. The 'Hundred Days' or second Napoleonic War which ensued after Napoleon's escape from Elba in 1815 caused a short-lived alarm but the defences were soon reduced to a care and maintenance status.

The records of the Napoleonic period contain some interesting domestic details of garrison life at Gravesend and Tilbury, and note that in 1807 the officers' barracks at New Tavern Fort received the luxury of wallpaper. In the previous year a plantation of walnut trees had been started in the fort grounds. It was not uncommon for such trees to be planted on suitable government property to provide, when mature, the material for gun stocks.

It should also be noted that in 1805 a very extensive royal dockyard was proposed to be sited on the marshes between Northfleet and Greenhithe and some land for it was purchased by the Crown in 1808. Had it been built, the plans called for fortifications several miles long on the bastioned trace, protecting it on its landward side.

A scheme which was undertaken was the 6-mile long Gravesend-Strood Canal, much of which was cut during the Napoleonic Wars. This like the earlier intended Gravesend-Tilbury tunnel, had both a military and commercial value but it was not completed until 1824.

As used at New Tavern Fort in the 1780s

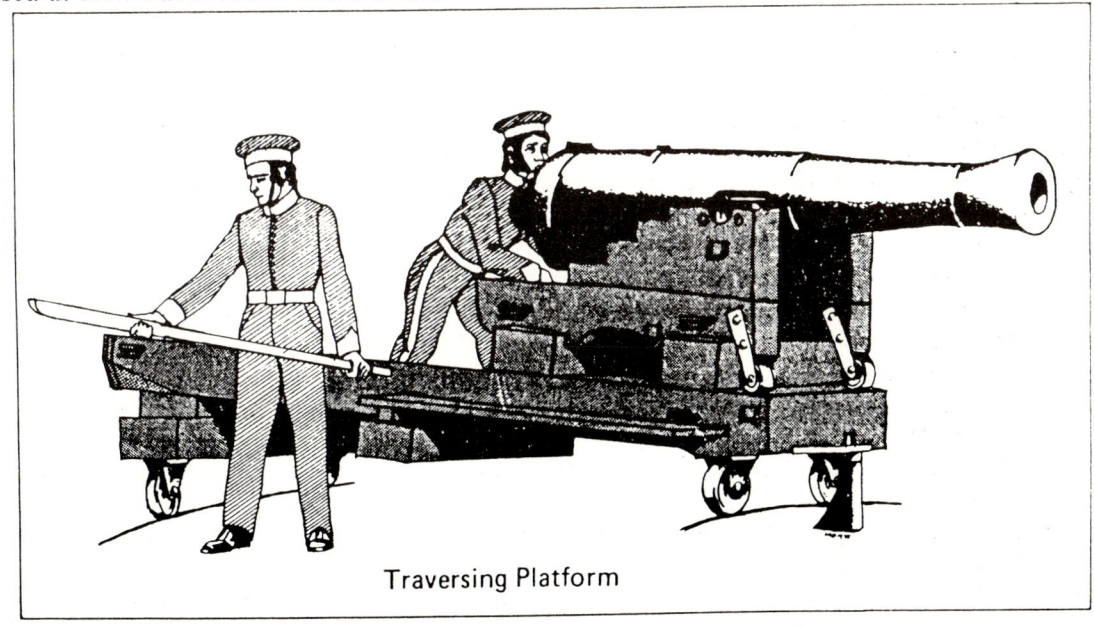

As used as New Tavern Fort in the 1840s (David Barnes)

The Revival of France

While post-war economy was attended by a reduction in the status of defences, to 'care and maintenance', it is quite erroneous to believe that France did not again become a menace in English minds until the 'Palmerstonian' scare of the late 1850s. Apparently contained by the terms of the Quadruple Alliance, France showed amazing powers of recovery and felt able to invade Spain in 1823 to restore Ferdinand VII. The policies of the Austrian Chancellor Metternich in the several decades following 1815 may have helped save Europe from a general war but they did not end British fears of France. Irrespective of whether or not France had any real capability to mount an invasion of these islands, there were certainly further invasion fears: in 1825 following the accession of Charles X, in 1830 on the accession of Louis Phillipe. Again, in 1848 it was feared that France might adopt an aggressive foreign policy as a means of diverting attention from her difficult internal problems — which led that year to a revolution and then the coup of Napoleon III.

The various scares led to the building of modest new works and upgrading of old ones in different parts of Britain but the Thames itself was excluded from this activity until the end of the 1840s. In the interim, the Gravesend blockhouse had been abandoned as a defensive work while the three forward batteries had been evacuated at the end of the Napoleonic Wars. However, in the 1840s, technical improvements at sea combined with growing fears of France led to a programme of defence construction in which the Thames shared. In 1844 the Duke of Wellington had drawn attention to Britain's vulnerability to the new steam warships which for short cross-channel voyages were not dependent on wind and tide and he proposed a programme of improvement to the coastal defences. Two years later, Palmerston produced a report along the same lines. Finally the government was stung into action as part of the defence programme which followed, money was found for the Thames.

At New Tavern and Tilbury forts, where new emplacements for 32-prd. guns on traversing platforms replaced the older emplacements on the same sites, the basic plans of the two forts being left unaltered. However, some new buildings were added at New Tavern, such as a large bomb-proof magazine, wash house, coal store and a guardroom. During excavations for the magazine, in 1848, some skeletons were found and it seems reasonable to presume that they were burials from the medieval Milton Chantry.

As to the forward defences, the battery at Lower Hope was left abandoned but at Shornemead and Coalhouse Point entirely new works were undertaken and a new school of defensive thought, the polygonal system was introduced into the Thames.

The Polygonal System and the second Shornemead Fort, 1847-52

Polygonal fortification grew out of a disillusionment with the bastion system which concentrated too much fire-power in purely flanking defence to the detriment of an effective frontal fire. The basis of the new system was simplicity and a judicious separation of the guns required for immediate self-defence from those to be used for long-range fire: the bastion was omitted altogether and a trace formed of straight lines of rampart and ditch which could be flanked from bomb-proof galleries called caponiers which projected into it from the lower part of the escarp. All the guns behind the rampart were available for frontal fire and the result was a more efficient and effective system than had existed in bastion for fortification. The origins of polygonal fortification can be traced in the treatises of a Frenchman, Montalembert in the late eighteenth century and in the designs for fortifications produced by Frederick the Great in Prussia. Even in the French Revolutionary and Napoleonic Wars, elements of the new system could be seen in England — in the Spur caponiers at Dover Castle and the ditch flanking casemates at Fort Clarence, Rochester. The second Shornemead Fort was probably the first pure example of polygonal fortification in this country.

The fort was pentagonal in plan, the three sides facing the river containing the armament, while the two landward facing sides formed the barracks for the garrison. The whole work was surrounded by an unrevetted ditch, crossed at the rear of the fort by a drawbridge. The armament comprised 13 x 32-prd. cannon with a range of nearly 1½ miles. They were mounted on wooden traversing platforms and fired through embrasures in the earthen parapet. Behind the guns on the parade was the main magazine with attached side arms shed, flanked by a concrete artillery store and an expense magazine. The barracks were of brick and were provided with loopholes to assist with the defence of the rear ditch, into which two open-topped musketry galleries projected. The front ditch was defended from two caponiers. Although employed in a river defence context this basic approach formed the prototypical design for English land fortification during the next 30 years.

In comparison with Shornemead, the battery at Coalhouse, finished by 1855, was on a most irregular plan. This was partly to be accounted for by the pecularities of the site — including the river flood defence wall which passed through it — and the retention of the earlier semi-circular battery within the perimeter. The fort mounted 17 guns located both in the old battery and in the river-facing parts of the new rampart. The rear was on a demi-bastioned line and contained the barracks which were arranged for musketry defence against a landing party. A broad water-filled ditch surrounded the whole work and the single entrance to the fort was located at the rear.

Difficulties were experienced in the construction of both forts. The contractor at Coalhouse required to be constantly supervised and the marshy nature of the ground produced uneven settlement at both works. Settlement problems were worse at Shornemead where cracks appeared in various parts of the brick superstructure, the cookhouse was in danger of collapse and one caponier had even completely detached itself. The records note that part of the bottom of the ditch was forced upwards from the thrust exerted by the weight of building materials piled on the ground nearby. However, by 1852 the fort was apparently ready for action although repairs continued to be necessary until 1861.

SHORNEMEAD FORT, 1855

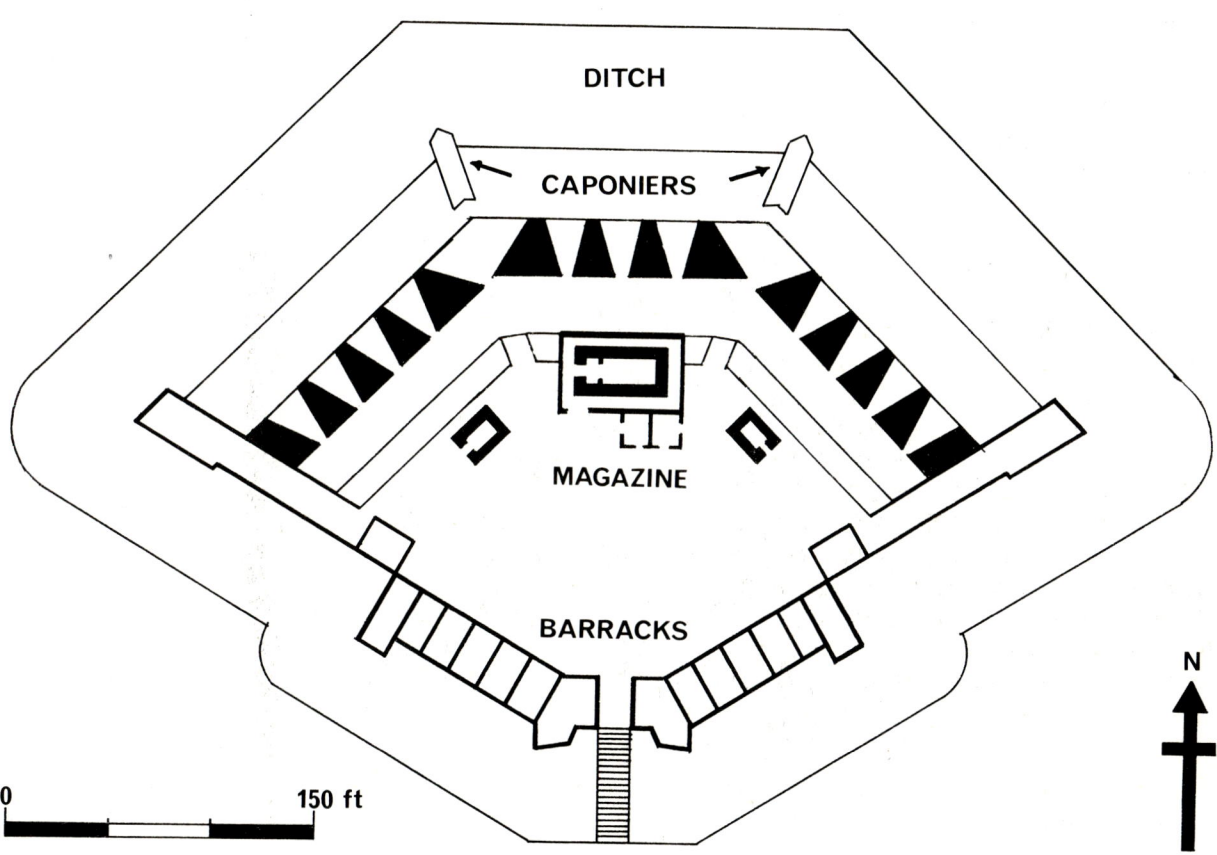

Technological developments and the Report of the Royal Commission on the Defence of the United Kingdom, 1860

Even as the new batteries at Shornemead and at Coalhouse Point had been under construction, Europe was experiencing a transition in the technology and technique of war which resulted in a 'military revolution' in the middle years of the century. This had many aspects but the most immediately relevant for coast defence were, firstly, the threat posed by the development of powerful rifled ordnance which seemed likely to supersede the smooth-bore gun and, secondly, the development of the steam ironclad. These ironclads had more efficient engines than the earlier steamships and armour was soon improved to become resistant to the existing smooth-bore guns firing roundshot or even spherical shell. Both of these developments were promoted by the considerable advances in industrial technology and manufacturing technique which had occurred since the 1840s. Some early signs of the possibilities for the future were seen during the Crimean War — in the use of the rifled gun and in the deployment of the French armoured floating batteries at Kinburn which anticipated the later ironclads.

The signs of things to come were to be demonstrated more forcefully in the American Civil War (1861-5) in which all the elements of the military revolution were deployed in this first of the 'modern' wars. Of particular relevance to British defence planners was the ability of the new rifled guns to reduce a brick and masonry fort to a heap of rubble (as seen in the bombardment of Fort Pulaski by the Federals in 1862) and the seeming invulnerability of armoured vessels to bombardment (as evidenced in the action between the Union Monitor and Confederate Merrimack in Hampton Roads also in 1862). Even in the later 1850s technical change — in particular the prospect of powerful rifled guns in the hands of the Continental powers and the more widespread use of steam propulsion in ships — would of itself have justified a revision of the nation's defences. However, technical change coincided with political circumstances which underlined that need: French aggrandisement; the apparent challenge to Britain represented by the French construction of the Suez Canal; suspicions of the ambitions of Napoleon III and his programme of construction of the very ships and guns which Britain feared, all helped foment a fear of war with France and invasion. The fact that Orsini's assassination attempt against Napoleon in 1858 had been hatched in England only served to heighten tension. The situation led in 1858 to a report on the effectiveness of the defences of the United Kingdom and to the raising of large numbers of army volunteers. The report had little to say about improvements to the Thames defences, apart from recommending a new land front at Tilbury Fort. It did however add that 'floating defences may ... be applied ... with great advantage' and suggested that 'it would also be a wise precaution to moor in the river some vessels with plugs in their bottoms, ready to be sunk at any moment for the purpose of stopping the passage in case of great emergency'.

A second report by a Royal Commission on the Defence of the United Kingdom in 1860, which instigated the most expensive scheme of defence yet in British history, considered and made recommendations on the Thames in a much more thorough and expansive way:

> 'The Defence of the Thames involves interests of vast magnitude; it includes the security of the great powder magazine establishment at Purfleet; the important arsenal at Woolwich and the adjoining dockyard; the Government victualling stores and ship-building yard at Deptford; the large amount of valuable property extending for many miles on either bank of the river; the fleet of merchant shipping moored in the port of London; and, lastly, the the metropolis itself.'

The Commissioners were of the opinion that although the positions of the existing forts were well selected, the works were 'insufficient to meet the description of attack that would probably be brought against them. The extent of injury that could be inflicted by an enemy who had succeeded in forcing his way up the Thames, renders it probable that a very powerful naval force would be employed in such a service.' The proposals of the Commissioners were as follows:

> 'We consider that the part of the river between Coalhouse Point and the opposite bank, where it is about 1,000 yards broad, is that best adapted for preventing, by means of permanent works, the further advance of a hostile fleet; and it has the advantage of being in immediate connexion with the lines which we propose for the land defence of Chatham on its western side, the right flank of which rests on the Thames at that spot. We recommend that the Shornemead Battery, which is admirable situated, should be enlarged, and, as its importance is considerably increased by its connexion with the proposed defences of Chatham, it should be converted into a strong work on the land side. At Coalhouse Point, on the left bank, a powerful battery should be placed in addition to or in extension

of the existing one, bringing the principal part of its fire to bear down the river and across the channel, but having some guns also bearing up the river in the direction of Gravesend. In addition to these, a work should be constructed on the right bank, opposite Coalhouse Point, at the southern point of the entrance to Cliffe Creek; and a floating barrier should be moored in time of war across the river, under the protection of these batteries, leaving a passage for our own vessels, for closing which every possible precaution should be taken at a time of expected attack.

In the event of the enemy's ships succeeding in forcing this first line of defence, in effecting which it is probable that he would receive considerable damage, he would then come under the fire of the batteries at Tilbury Fort and Gravesend; and we consider this second line so important that we recommend that these works should be put into the most thoroughly efficient state in every respect; their guns would cross their fire, at a distance of 2,000 yards, with those on Coalhouse Point and Shornemead; and a similar obstruction or floating barrier to that above recommended should be prepared, to be moored between Gravesend and Tilbury Fort.'

To link with the defences of the Medway and to help prevent a landing in the Slough Point area, the Commissioners also proposed a fort at Allhallows. A proposal to build a line of land-forts from Shornemead to Telegraph Hill at Strood was not proceeded with.

It was reasoned that to have begun all this work at the same time would have left the Thames defenceless and so it was decided to leave New Tavern and Tilbury Forts until the first-line works at Shornemead, Cliffe and Coalhouse were complete.

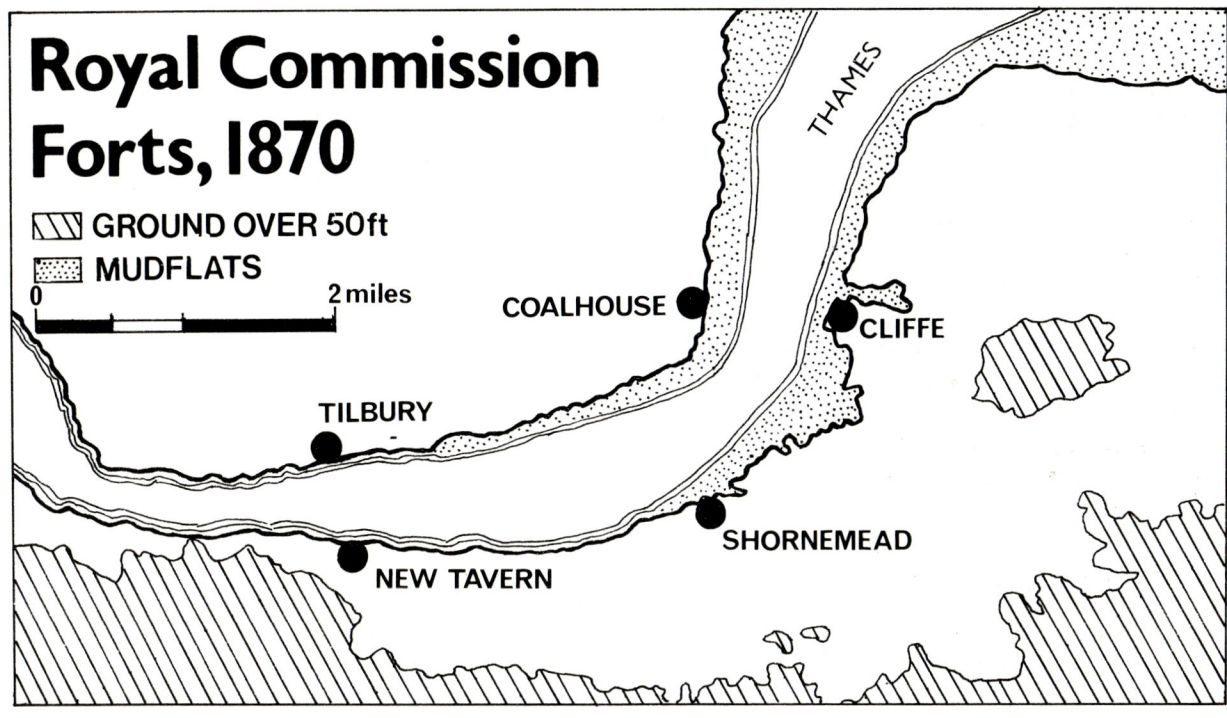

The First Line Defences

These were commenced in 1861-2 and took over ten years to build. Their progress was dogged by structural problems caused by the unstable and marshy subsoil, the difficulties being most acute at Shornemead as they had been in the 1840s. Living conditions for those working on the forts was not pleasant, the supervising officers often suffering from bronchitis and fevers. As conceived in 1860, the forts were to be two-tiered works, consisting of a range of gun casemates protected by iron shields, surmounted by guns placed on the roof. Although construction actually began on this basis, even as work was underway the design of the works was several times reconsidered, notably in 1867 and 1869. One proposal was for alternating gun casemates with pits for Moncrieff disappearing guns all on a single level but in the end the decision was more or less the original plan of 1860, minus the guns on the roof.

By the early 1870s the works were complete. Each consisted of an arc of granite-faced gun casemates with iron shields and an open battery at the up-river end with the rear closed by a barrack of Kentish Rag. The casemates received heavy protection against high angle fire, in the form of a 5-ft. thick roof of concrete and brick, internally formed into compound vaulting. All the guns were serviced by magazines directly below them at basement level, and communicating with them via ammunition lifts. The forts were designed to be self-defensible and so their casemated fronts were protected by ditches containing musketry caponiers and the windows of the barracks had internal steel shutters pierced with loopholes for small arms. All-round defence was therefore possible.

When the Royal Commission had made its recommendations in 1860 the mainstay of British coastal defence was the 68-pdr., a smooth-bore with a range of about 3,000 yards. But this was useless against the armoured vessels mounting heavy rifled guns with which the Continental powers were equipping their navies in the 1860s. Therefore the forts had to be armed with weapons powerful enough to penetrate the thicker armour being carried on ships. When the guns for the forts were supplied from around 1874, they were the heaviest and most powerful available: the lightest of them, for the open batteries, was the 9-in. rifled muzzle-loader firing a 250-lb. shot to a range of about 4,500 yards; the casemates were armed with 11-in. guns firing 548-lb. shot and 12.5-in. firing an 820-lb. shot. These giant guns, weighing up to 38 tons, were placed on an updated and all-metal version of the earlier traversing platform which was provided with all the hand-gearing for traversing and elevating which Victorian engineering could devise. With these powerful weapons it was not sufficient to merely incline the plane of the platform to take up recoil — a special device had to be introduced to help absorb the forces involved. At first, compressor plates which operated like one comb moving through another slightly tighter comb were introduced. But within a few years this was replaced by a hydraulic piston.

In some of the still surviving casemates — particularly those at Coalhouse Fort — the metal rails or racers on which the guns traversed can still be seen, together with such details as the eye-bolts used for the tackle to mount and dismount guns, the bars which held the rope mantlets or mats hung around the gun at the port to keep bullets and splinters from entering the casemates during an attack, all within some impressive brick vaulting.

Front of gun casemate at Coalhouse Fort (V.T.C. Smith)

Entrance and defensible barracks, Coalhouse Fort (V.T.C. Smith)

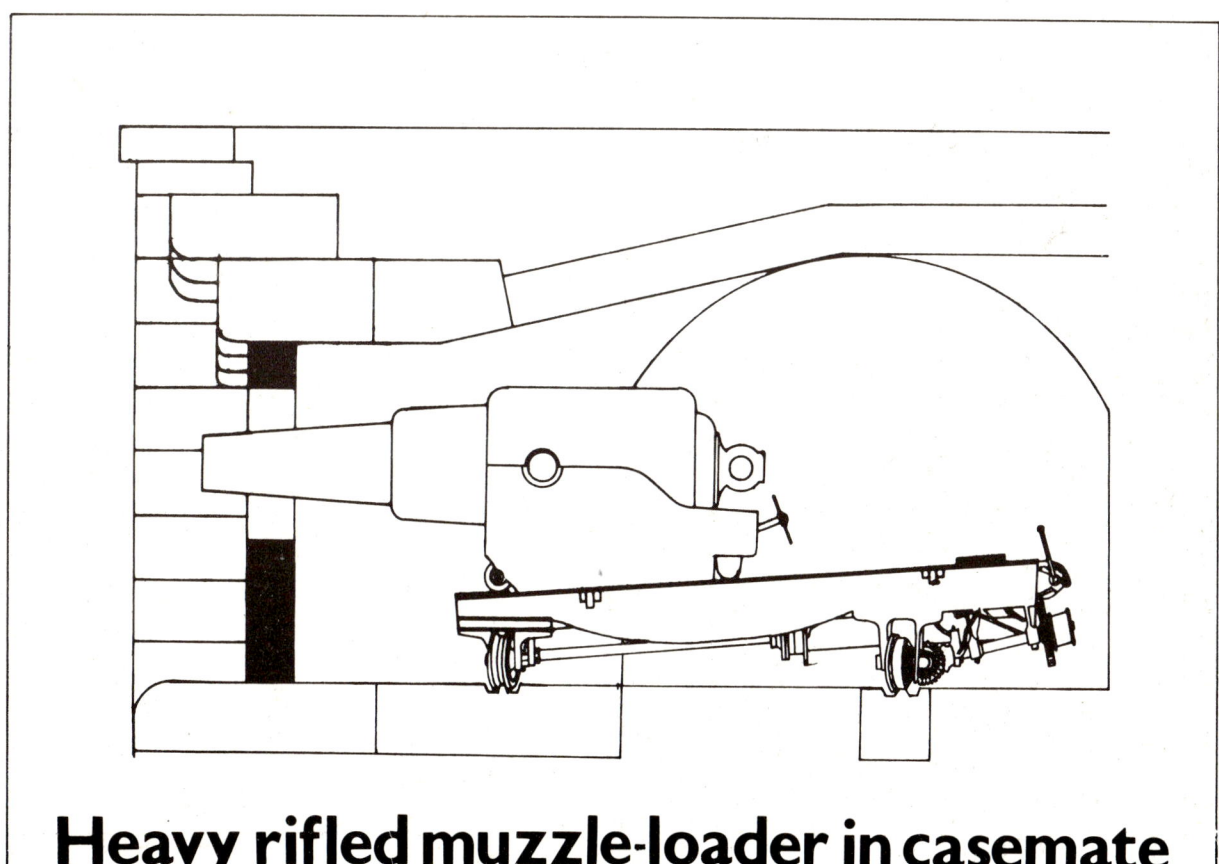
Heavy rifled muzzle-loader in casemate

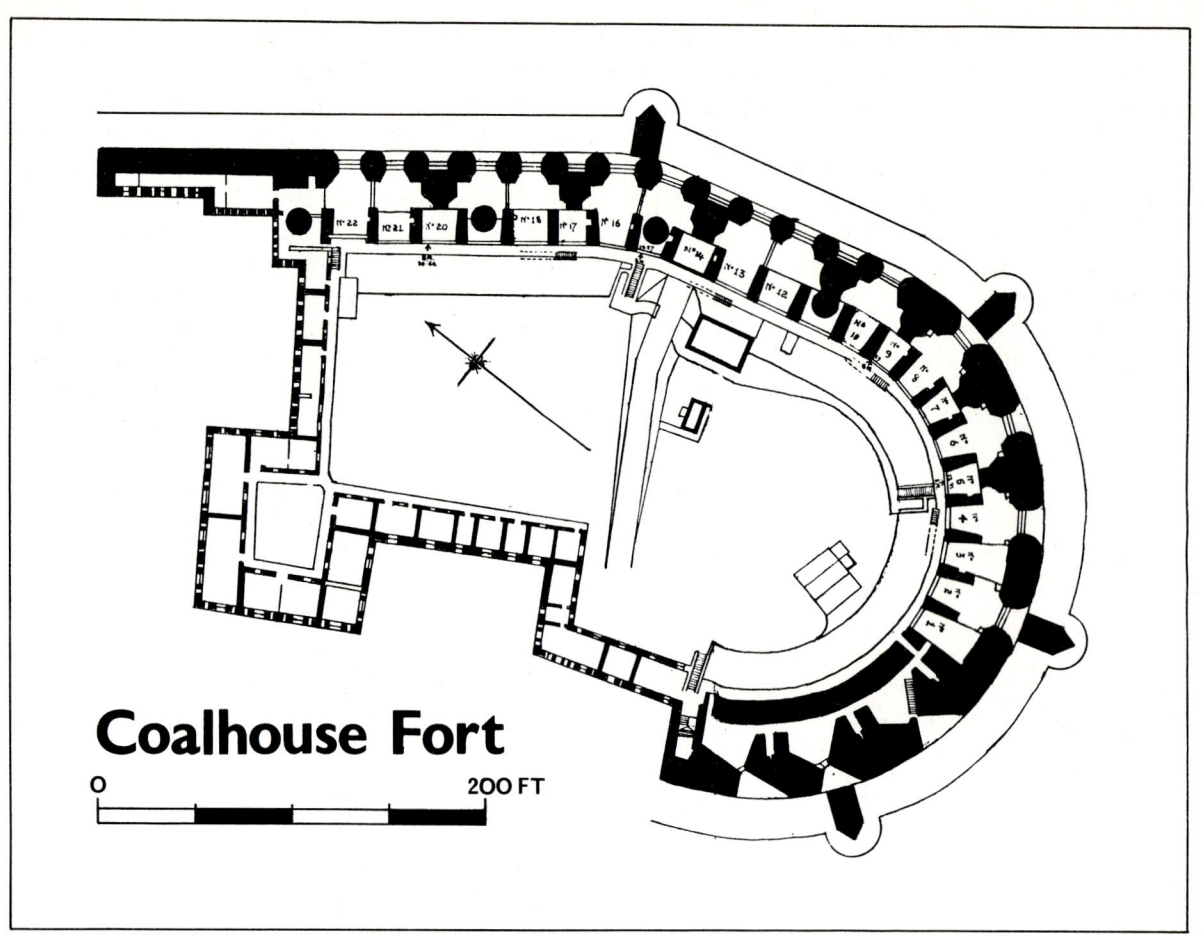

Magazine Lantern (*National Army Museum*)

Just as artillery had made rapid advances, so the arrangements for ammunition storage and handling had been improved. In place of the old arrangement of detached magazines out in the open, wherever possible magazines were now buried deep under the emplacements or rampart for protection and had direct communication, via lift shafts, with the emplacements above. Special precautions were taken in the lighting of magazines to ensure that no naked flame was taken through an area where explosives were present. Therefore illumination was effected externally — the individual magazine chambers were lit by lamps placed into glass-fronted recesses from the other side of their walls and reached from a lighting passage. The ammunition passage was illuminated from globes in the roof, also served from the outside. Metals likely to strike sparks were so far as possible avoided in the magazines — even down to hinges and screws. In addition, magazine chambers were lined with timber battens to prevent shells and cartridge cylinders knocking grit out from the walls which might be struck and cause sparks. All magazine workers were stopped by a barrier at the entrance to remind them to remove dangerous hobnail boots and their uniforms which might contain particles of grit, before passing over to put on special magazine clothing.

Of these forts of the first line, only the facade of casemates survives at Shornemead. Cliffe is nearly complete but the barracks have been gutted. Coalhouse, the larger of the three, is in the best condition and is at present under restoration.

Heavy rifled muzzle-loading gun in casemate, as mounted in the Thames Forts, at Fort Delimara, Malta   *(Quentin Hughes)*

The second line — Tilbury and New Tavern Forts

Some improvements had already been carried out at these forts in the late 50s and early 60s. These included replacement of some of the guns with a heavier smooth-bore armament and the provision of an 18-pdr. field battery at both places. An old magazine in the SE bastion of Tilbury Fort was replaced and, possibly a new magazine was built at New Tavern Fort. In 1865 the armament of Tilbury Fort comprised 5 x 68-pdrs., 5 x 32-pdrs. and 4 x 10-in. shell guns in the gun lines as well as many other smooth-bores elsewhere in the fort; at New Tavern there were 8 x 68-pdrs., 2 x 10-in. shell guns and 1 x 8-in. shell gun. These had been in place since 1858. However, despite the increased fire-power, none of these guns was in the armour-piercing class and were merely a palliative while the forward batteries were under construction. In 1866, during an era of 'turret mania' in defence schemes it was actually proposed to install a gun turret on the Tilbury blockhouse but sensibly, the scheme was not proceeded with. However, in 1867 the monster Horsfall Gun of 13.3-in. calibre was mounted at Tilbury. This smooth-bore had been tested on the Shoeburyness range in 1866 and for a time it appeared that its fair degree of success might give a new lease of life to smooth-bore ordnance but it did not prevent the transition to rifled guns.

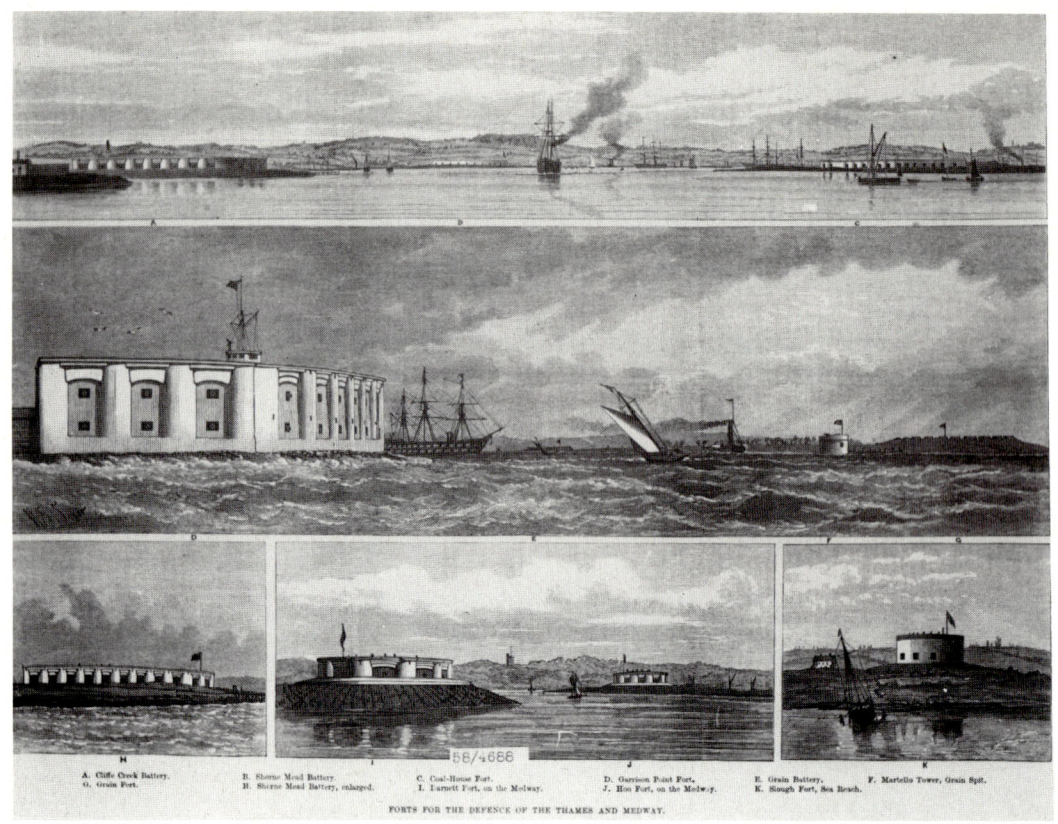

Views of several of the Thames and Medway forts built under the Royal Commission scheme
(Illustrated London News)

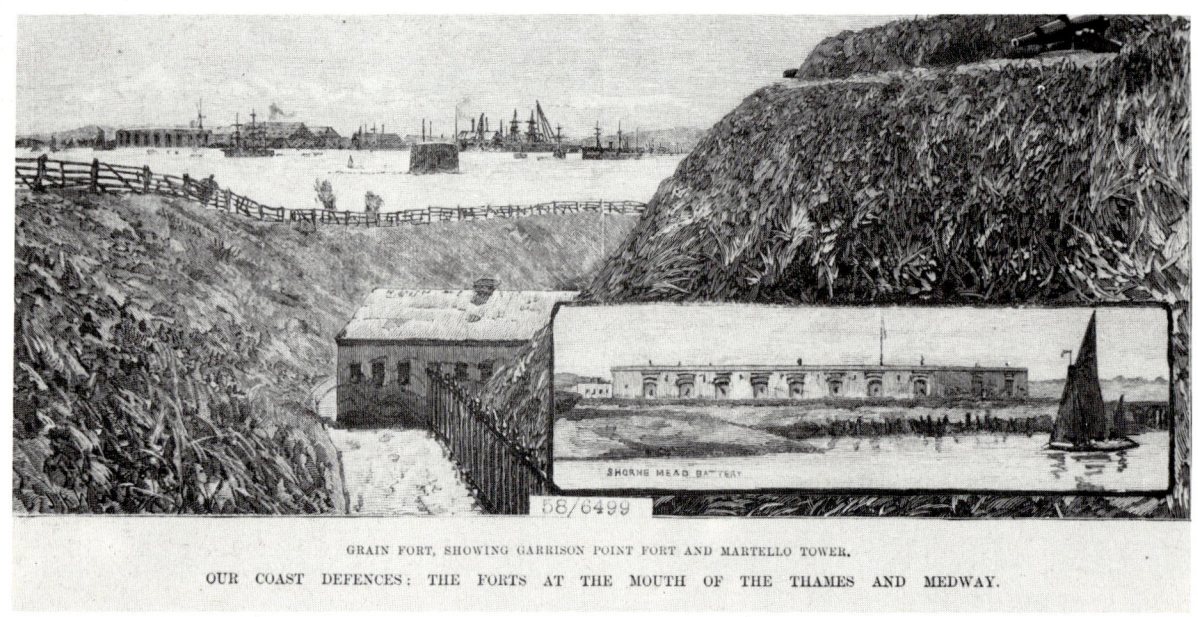

Grain Fort in the Medway, with Shornemead Fort inset (Illustrated London News)

Work on remodelling Tilbury and New Tavern under the Royal Commission scheme began in 1868. At Tilbury Fort, brick emplacements for 13 guns (by 1874, 1 x 12-in. RML and 12 x 9-in. RML) were built on the SW, NE and SE bastions and south curtain, but the basic plan of the fort was not altered by this. Unlike the closed casemates of the first line forts, these were all open to the sky but those in the SE bastions and south curtain contained the same type of massive wrought iron shields. An example of one of these shields can still be seen in rear view within the south curtain. The other emplacements of the curtain and SE bastion have been obscured by later work. The emplacements in the SE and NE bastions (unarmoured) still survive. All the emplacements were separated from each other by brick ammunition serving rooms into which the ammunition was raised from identically sized store rooms below. These stores, alternately for shell and cartridge, communicated via a passage with a main magazine. As with the first line forts, there was the same system of safe lighting. Sadly, the work at Tilbury involved the demolition of the old Tilbury blockhouse but its foundations may still exist under the south curtain rampart.

New Tavern Fort was improved with similar positions for 10 guns (1 x 12-in. RML and 9 x 9-in. RML), all but three of which were fitted with iron shields as at Tilbury. One of the armoured emplacements is being restored to its original appearance. Electrical detection surveys would suggest that several more shielded emplacements lie buried under later work. The magazines paralleled those at Tilbury Fort except that they contained spare cartridge and shell stores in addition to a main magazine. The magazines are currently being restored to their former state.

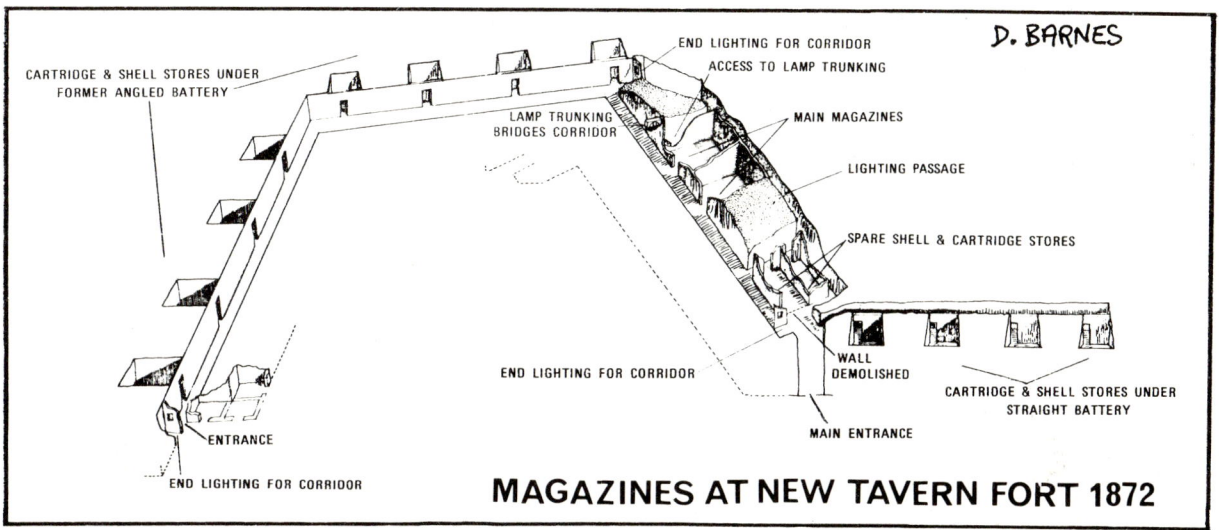

MAGAZINES AT NEW TAVERN FORT 1872

Slough Fort, which was like the first line forts but in miniature and with a semi-circular earthwork to the rear, had been completed by 1867. So far as can be judged however, it was never actually fitted with shields although they had certainly been intended.

New Tavern and Tilbury forts had been largely disarmed in 1868 prior to reconstruction. Since none of the Thames forts can be shown to have been provided with guns before 1874, the Thames had been practically without gun defence for the previous six years.

With steam ships able to cover perhaps a quarter of a mile in a minute, the number of rounds which could have been fired at them from the slow-firing guns of the defence was quite small by today's standards and with little chance of scoring a hit at anything but short range. Yet it is sometimes forgotten that the guns were only part of the mechanism of defence. Integral to the Thames was a system of boom obstacles and mines involving a whole new technology of submarine destruction. In addition to the fort guns and the mines, field guns and machine guns would be deployed in the neighbourhood to prevent a landing, while the marshes on both sides of the river could be flooded. It has been said that the river forts were vulnerable to bombardment from the commanding heights of the hinterland but those very heights would have been entrenched in wartime to prevent their use by an enemy.

Positioning of iron shield on front of experimental casemate at Shoeburyness, 1860s
(R.E. Corps Library)

Heavy rifled muzzle loading guns firing in simulated attack (Navy and Army Illustrated, 1899)

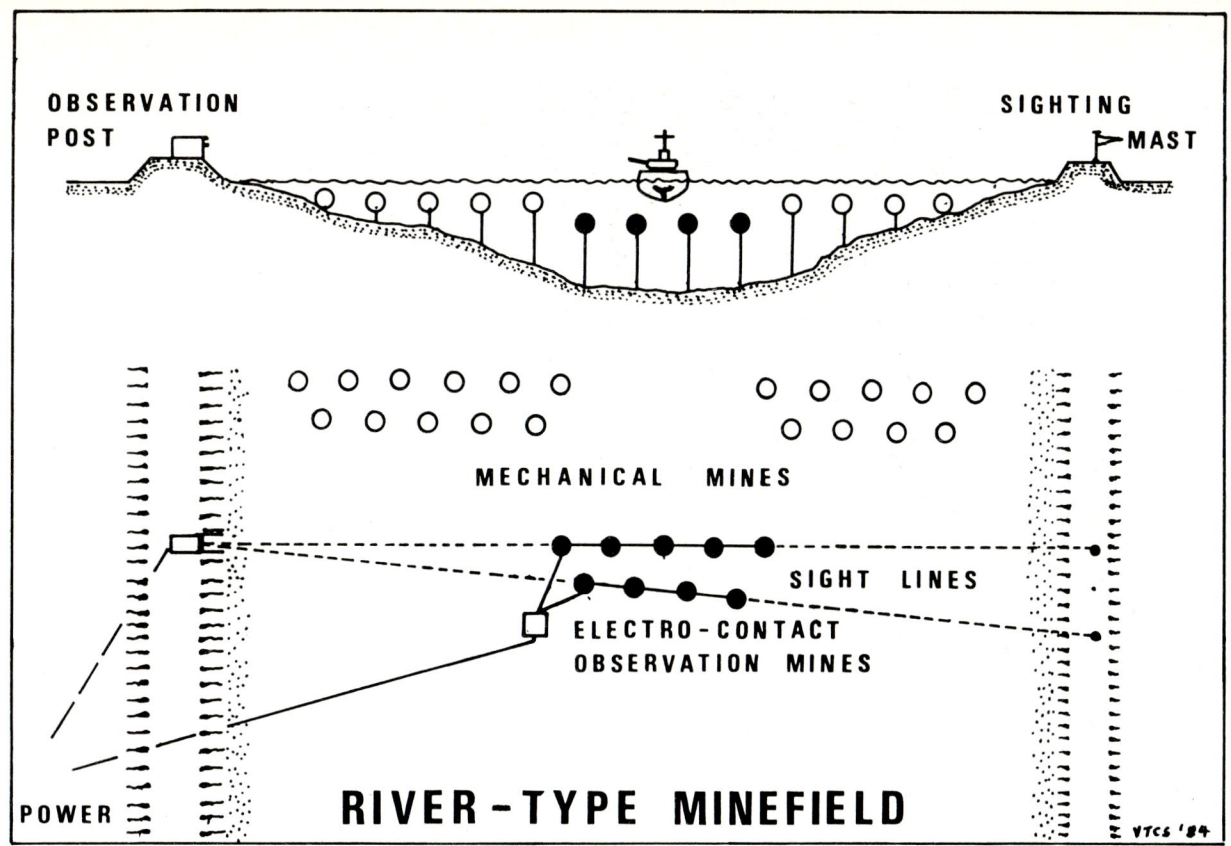

Later nineteenth century — the age of breech loading

Far from disappearing, after the defeat of France by Germany in 1870 and the demise of Napoleon III, the fear of a possible French invasion continued to assert itself throughout the remainder of the century. The threat of the combined fleets resulting from the alliance of France with some other power such as Russia, was particularly feared.

For the most part, the Royal Commission forts and armaments formed the basis of British coastal defence for the next 15 years. In the Thames, the only differences during this period were a few unradical changes in the calibres of the guns mounted. Yet if these were uneventful years for the Thames defences they were certainly very formative ones in the development of artillery and the science of fortification: the Prussian artillery in the war of 1870 had shown the superiority of breech-loading ordnance over muzzle-loaders while in the intervening years the ranges of guns and the destructive power of their ammunition had greatly increased. By the mid-1880s, the warships of the European powers were faster, more heavily armoured and provided with more powerful and longer range guns than those with which the Royal Commission forts had been expected to cope. Equally, the well-marked vertical expanses of masonry offered by the casemated style of fortification and the close packing of guns — so vulnerable to cumulative explosion — were a liability in the face of the new and more accurate weapons, especially the powerful breech-loaders. If the British defences in general and the Thames defences in particular were to be viable, they had to be improved both in armament and in design to counter changing circumstances.

In the forward defences of the Thames, the initial response was a proposal to adapt the existing and out of date system by removing guns from alternate casemates which would then be filled in and their armour added to the fronts of the casemates in which the guns were retained. This would have ended gun-packing and the danger of cumulative explosion and would have given the remaining guns twice the armour protection. However, the price to be paid would have been a net decrease in fire-power. Another proposal was to build a massive glacis of chilled iron blocks on to the fronts of the casemates to give them dramatically increased protection. This very expensive approach was not adopted although a few casemates at Coalhouse and Cliffe were filled in with concrete to provide traverses.

In about 1888, an additional defence was added in the form of a Brennan Torpedo installation at Cliffe Fort, well sited to hit vessels as they slowed down to make the turn into Gravesend Reach. The Brennan dirigible torpedo was the invention of an Irish-born Australian, Louis Brennan and was officially adopted in 1887 by the War Office as a harbour defence weapon, being subsequently deployed not only at Cliffe but at Garrison Point in the Medway and at various other places in the

UK and in the colonies. The Brennan Torpedo was actually powered from the land on which a steam winch unwound two drums of piano wire in the hull, linked to two contrarotating propellors which forced the weapon through the water at a depth of 8-10 feet and at a speed of up to 20 miles per hour. With a warhead of 200 lbs of guncotton, it was a formidable device and could be delivered to target with considerable accuracy. The same wires which powered the propellors could be used to steer it from the land, a flexible mast protruding above the water to allow tracking by the shore-based operator. It was claimed to be able to hit an object as small as a floating fruit-basket at 2,000 yards and to be capable of being turned through 180 degrees and so attack a ship from the offside. At Cliffe two launching bays, and the rails down which the torpedoes were launched still survive, together with the circular high placed directing station (the site of a telescopic conning tower) and underneath, the engine room and torpedo storage area. Apertures to take the wires from the engine room to the torpedoes can still be seen.

Brennan Torpedo *(R.E. Corps Museum)*

Eastern launching bay, Brennan Torpedo Station at Cliffe Fort *(V.T.C. Smith)*

The Brennan was only one of a number of different systems available from different inventors e.g. the Lay, Patrick, Berdan, Ericcson, Sims-Edison, Nordenfeldt and Whitehead torpedoes, either controlled or free-runners and using differing forms of motive power. The Whitehead was the torpedo of the future but at this date was considered inferior to the Brennan. The Brennan torpedo was a close-attack weapon, an adjunct, rather than a real defence in itself and merely shored-up a basically out of date system of short-range, slow-firing RMLs.

The need was for fundamental change and this arrived in the form of the faster-firing breech-loading gun mounted on a disappearing carriage, an updated version of the earlier Moncrieff carriage. In this weapon system the architectural monumentalism of earlier years with its stark vertical faces, as in the casemate forts, was rejected in favour of very low-profile structures hardly protruding at all above ground level. In an effort to reduce the target area, the gun was mounted in a pit and only rose up above the parapet at the very moment of firing and then recoiled back down again for reloading. Unlike the older Moncrieff arrangement which had used a counterweight, this system employed hydraulics: the force of recoil on firing, compressed oil in a recuperating cylinder linked to an arm to raise the gun. When the pressure was suddenly released by a trigger, a piston forced the gun up above the parapet for firing and so the process was repeated.

Total concealment was thus achieved broken only for a few seconds when the gun fired. These weapons had a far greater range than the old and clumsy rifled muzzle-loaders — in excess of 8,000 yards — vastly improved traverse and a significantly higher rate of fire. Their security against bombardment lay not in massive armoured protection but in their invisibility, for a target which cannot be seen is very difficult to hit.

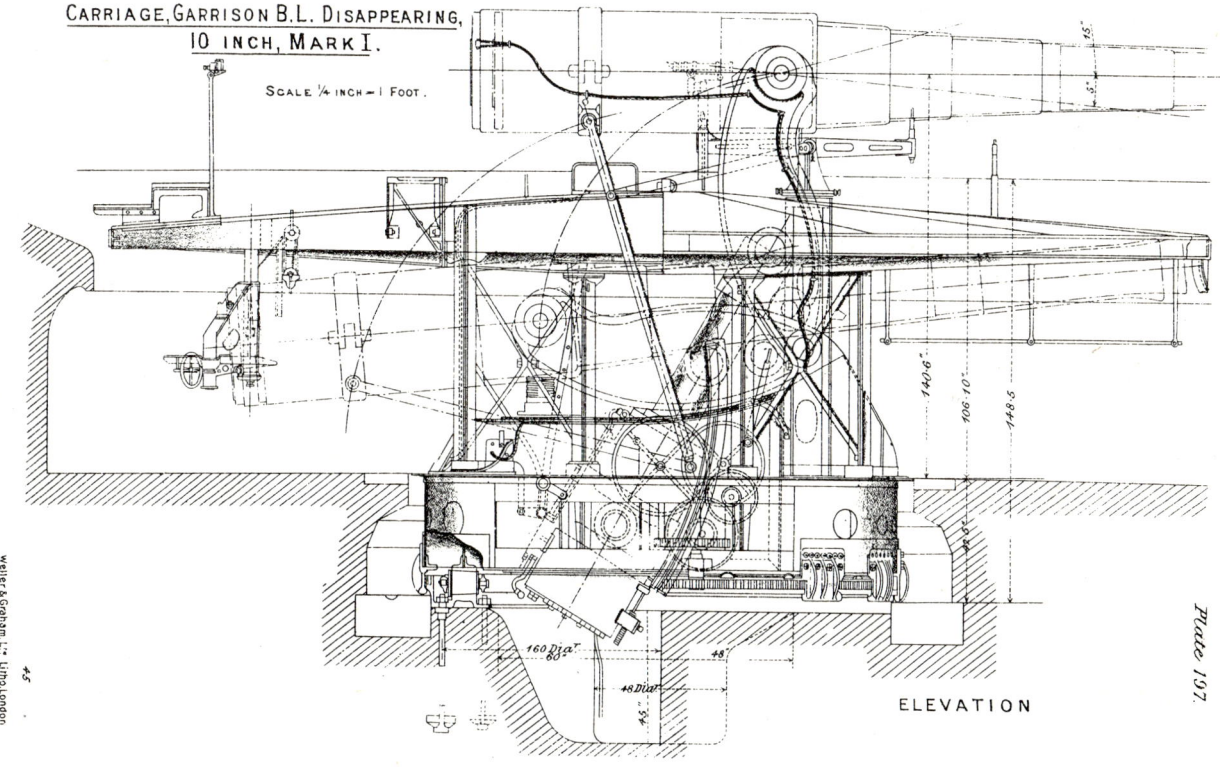

(Treatise on Carriages, 1911)

Such batteries were built in a number of sites both in England and abroad. In the Thames two were built; at East Tilbury, on the crest of gently sloping ground about 500 yards to the north of Coalhouse Fort (1891-2) and at Slough Fort (1895). East Tilbury Battery was equipped with 2 x 10-in. and 4 x 6-in. guns, with a 3-pdr. quick-firer on the right flank, probably for practice. This battery was very carefully blended into the landscape and was surrounded by a shallow ditch containing an unclimbable fence concealed at the foot of the glacis. The slight profile of this work and the position of the ditch and fence at the extremity of the glacis was copied from the Twydall profile which had been invented during the construction of the Chatham land defences a few years earlier. Although later modified and now overgrown, East Tilbury battery is still in a reasonable state of preservation. Its deeply placed magazines, their interior walls rendered with bitumen and cork, still retain their lift equipment for raising the shell and cartridge to the guns.

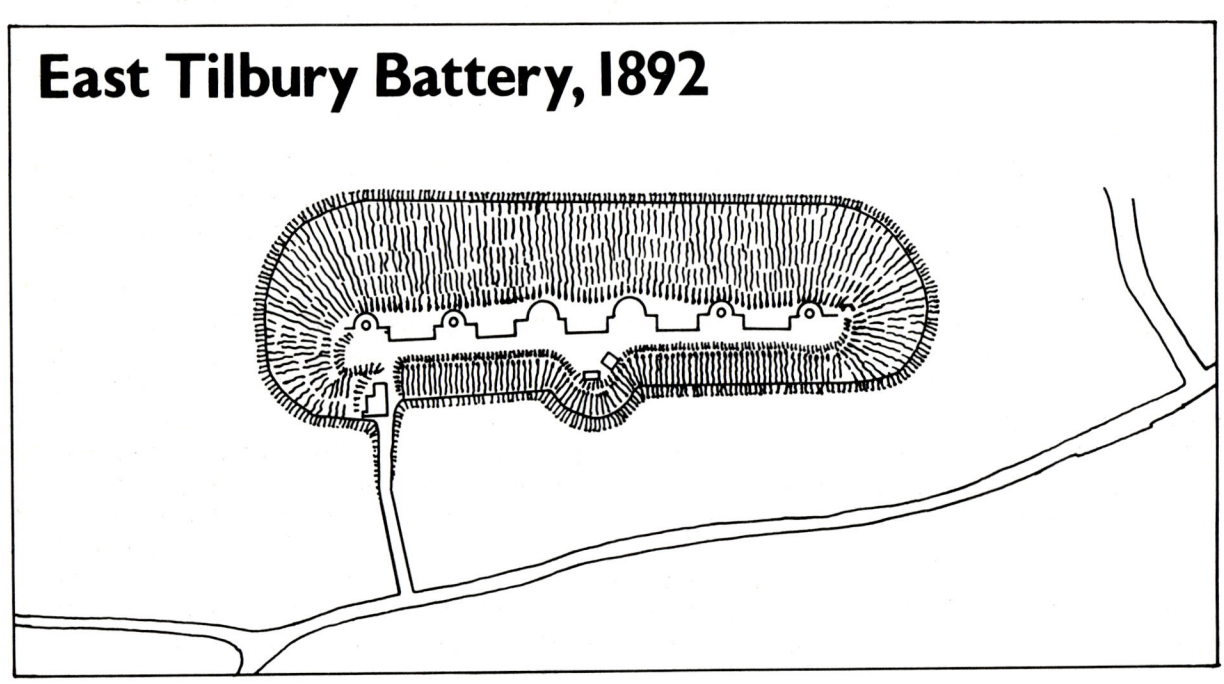

10 inch breech-loader on a disappearing carriage, similar to the weapons mounted at East Tilbury battery   *(Royal Artillery Institution)*

East Tilbury Battery, 1892

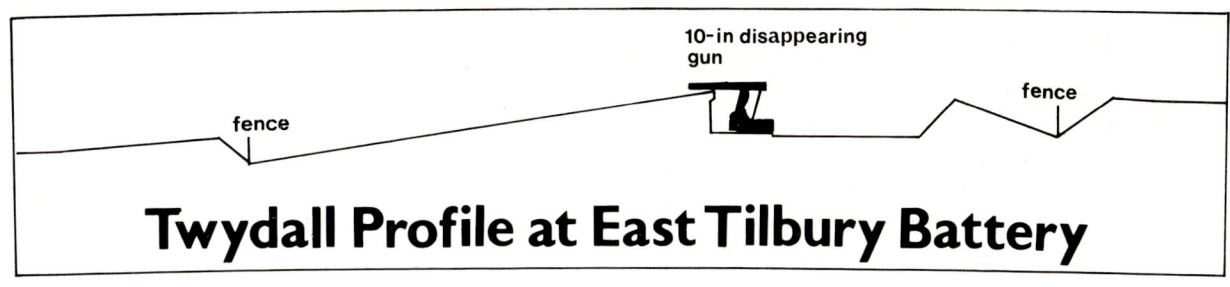

Twydall Profile at East Tilbury Battery

East Tilbury Battery from the air (V.T.C. Smith)

The battery at Slough was built as 'wings' on either side of the fort: the right wing was for 2 x 9.2-in. and the left wing for 2 x 6-in. The right flank remains, but it has been reconstructed for other guns and although some traces of the unaltered left flank may be seen, most of it has been filled in.

These long range guns, initially supplemented but did not immediately replace the RMLs. The provision of the Brennan torpedo station at Cliffe in the second half of the 1880s had in no way superseded the earlier submarine mining arrangements. Rather these were elaborated upon and made subject to greater experimentation and training efforts, Shornemead Fort becoming a training centre for the Thames and Medway submarine miners. Each summer in the closing years of the nineteenth century and in the first decade of the twentieth, tugs were hired from Gravesend to practice mine laying in the river. Mines and boom defences had become more important than ever, especially against the feared predatory incursions of fast torpedo boats with which France and several other Continental states had begun to equip themselves. This class of vessel had become an increasing menace since the 1870s. It was feared that the great speed and manoeuvrability of torpedo boats would enable them to carry out lightning raids on shipping in British military harbours and anchorages with comparative immunity against the existing slow-firing medium and heavy defence guns.

British torpedo boat steaming at high speed, c 1890 (Navy and Army Illustrated)

The danger was underlined by the example of the highly successful raid in 1895 by Japanese torpedo boats on Chinese warships in the harbour of Wei-hai-Wei (defended by guns and a boom), during the Sino-Japanese War. The need to counter the speed of the torpedo boat and to reinforce mine and boom obstacles gave rise to a new class of land based gun — the light quick firer (Q.F.). It was initially of small calibre — 3-pdr. (firing 30 rounds per minute) — but was later developed into the heavier 6-pdr. (25 rounds per minute) and, finally, the 12-pdr. to cope with the increasing size of torpedo boats and the protection given to them. At first, the shortage of Q.F. guns in Britain was responded to by the issue of special case shot (filled with 3-lb. balls) for the RMLs. However, by the mid-1890s, 3 x 3-pdr. Hotchkiss or Nordenfeldt QF guns had replaced the RMLs in the open emplacements at Cliffe Fort and a detached battery of 4 x 6-pdr. was built a few hundred yards to the right of Coalhouse Fort. A further detached battery for 2 x 6-pdr. was added a few yards downstream of Shornemead Fort. All these guns were supported by adjacent Defence Electric Light emplacements to illuminate the river for night firing.

Electric light emplacements on the roof of Coalhouse Fort; Second World War observation post in foreground (V.T.C. Smith)

The defensive arrangements were tested in September 1895 during the annual mobilisation of the Royal Artillery to man the Thames forts. The forts had to be prepared against an attack by torpedo boats from the squadron at Sheerness. The operations seem to have been more instructive than successful: the mists to which the river was subject were clearly a source of insecurity for the electric lights had proved incapable of penetrating them during the mock attack. Had the torpedo boats chanced a dash through they would have been quite unseen. The exercise showed that in the procedures for transmitting signals from observing stations to the forts there was room for improvements. At the end of the exercise the disappearing guns at East Tilbury Battery were live fired and a Brennan torpedo from Cliffe Fort was successfully run.

The mixture of old RMLs and a few heavy BLs, with a collection of light QFs in the Thames in the mid 1890s was generally representative of the situation everywhere in British coastal defences at that time. The RMLs had been obsolete against the heavy breech-loading guns in the navies of the Continental powers for well over a decade but as yet only faltering steps had been taken to effect the re-armament of our coastal forts. The designs and calibres were still in doubt as ship-board armaments improved. However, during the next ten years most of the RMLs were withdrawn from the Thames forts as elsewhere until there was total reliance on BL equipments. The increased and increasing ranges of the weapons involved (by 1905, seven or more miles for medium and heavy guns being possible) made it possible to engage targets further and further from the object to be defended. This was very important since shipboard guns could fire at shore targets from a balancing further range. Thus it happened that by 1905, the heavy BL guns recently mounted at Grain Fort and Sheerness, places which in the RML era had been capable only of defending the entrance to the Medway itself, now became the first line of defence for both rivers since they could reach well out into the estuary. The remaining RMLs were gradually removed from the Thames forts and replaced by medium and light guns intended to block any attempts at penetration and to defend booms and minefields: 4 x 6-in. and 4 x 12-pdr. were mounted on the roof of Coalhouse Fort and

4 x 12-pdr. were added to Cliffe Fort whilst the nearby Hope Point Battery gained 2 x 12-pdr. Further upstream, Tilbury Fort was provided with 2 x 6-in. and 4 x 12-pdr. on the SE Bastion and S curtain while New Tavern Fort received just 2 x 6-in. although for a few years 3 x 9-in. RML were retained. In all cases this re-armament was attended by a certain amount of further earthing up of the fronts of the works and the planting of bushes and shrubs in an attempt to break up harsh outlines.

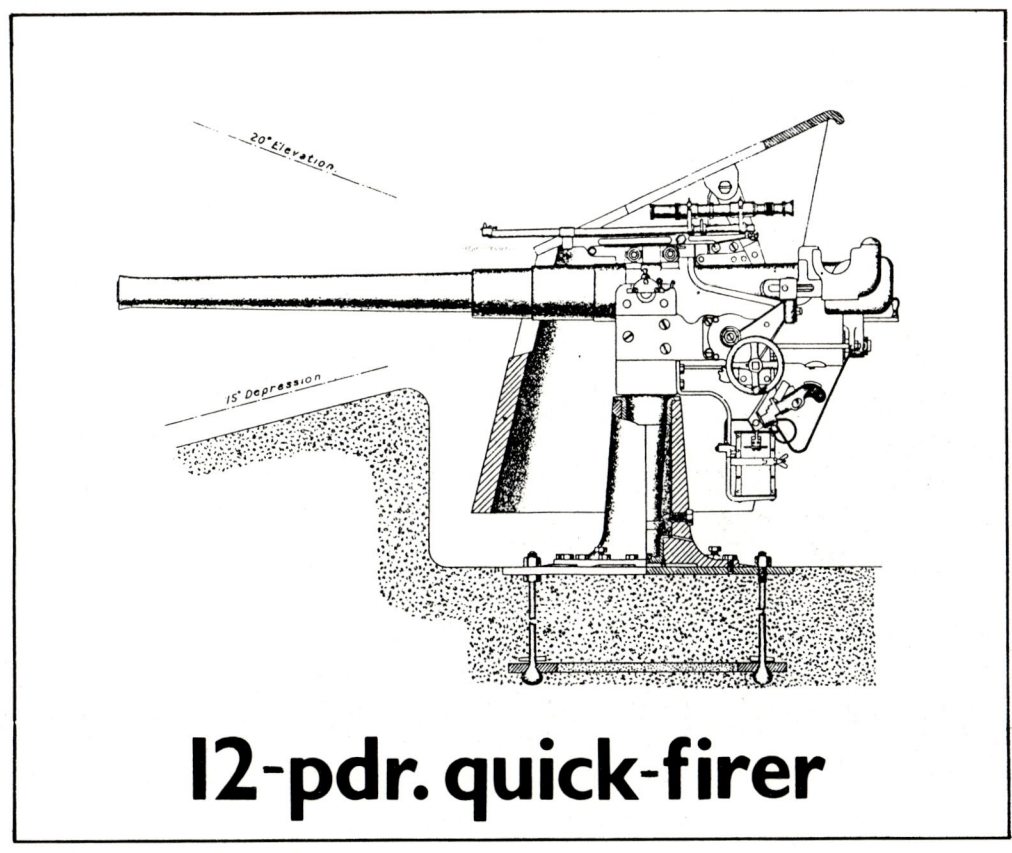

12-pdr. quick-firer

As mounted in the Thames forts (Manual of Carriages)

Diagram of range finding

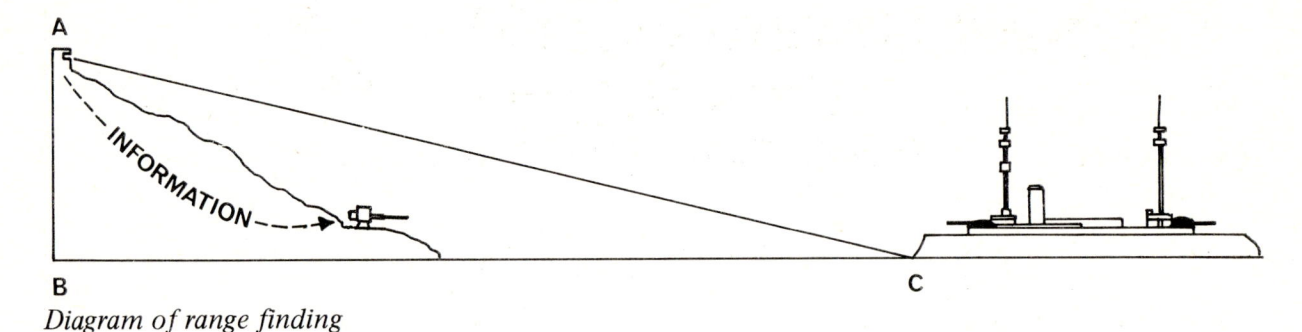

Not only had breech-loading been a radical departure from the past practice but so also was the advent of accurate range finding and effective fire control. Even in the heyday of the RML era, the accuracy of range estimation left much to be desired. In addition, the tangent sights then used were only efficient for relatively close-range targets. For the fort commander to control the fire of his guns there had been no uniform system, orders being passed by runner, megaphone, trumpet or by speaking tubes, all of which could be difficult in the noise and heat of battle. It was not until nearly the end of the RML era that an effective system of range finding and gun control was introduced. This was Captain H.S. Watkins' system which had been devised in the late 1870s and first introduced on an experimental basis in coast defence during the mid-later 1880s. The break-through came with Watkins' Depression Range Finder, fitted with a moveable telescope which an observer (A) sighted on to the target (C). By measuring the angle of depression (ACB) from the instrument whose height above the water (AB) was known at any tide state, the range (BC) was determined. This information was relayed by electric telegraph to the guns whose captains read it off an electric dial and laid their weapons accordingly. This gave the commander in his fire-control cell an unprecedented capability to both fire his guns accurately and to control their fire. The Depression Position Finder, another invention of Watkin, gave the gun layer both traversing angle and elevation. For close-range work the autosight was used. This fitted on to the guncradle. These instruments were generally introduced into British coast defences from the later 1880s and early '90s. In the Thames they were introduced at the end of the RML era at Coalhouse and Cliffe Forts where observation cells were constructed for them on the roof. There is also some evidence to suggest installations at Tilbury Fort and at Shornemead. However, all the surviving positions belong to the BL era. The best example is the one on the roof of Coalhouse Fort, which has an annexed room for the Electric Light Director.

By the opening years of the twentieth century then, the Thames was possessed of a modern system of defence. This was based upon heavy long-range breech-loading guns at East Tilbury and Allhallows with others at Grain and at Sheerness firing far into the estuary, backed by medium and light guns, minefields, booms and electric lights upstream. In addition, any guns on the Shoeburyness gunnery range which could be brought to bear were to be regarded as part of the Thames defences.

Heavy gun and transportation barge 'Magog' at Shoeburyness gunnery range, c 1895
(Navy and Army Illustrated)

Thames Defences, 1914

[Map showing Thames estuary with batteries marked at Canvey, Tilbury, Gravesend, Sheerness, Shoebury area, with boom near Sheerness. Chatham and mudflats also indicated. Scale: 0–5 miles.]

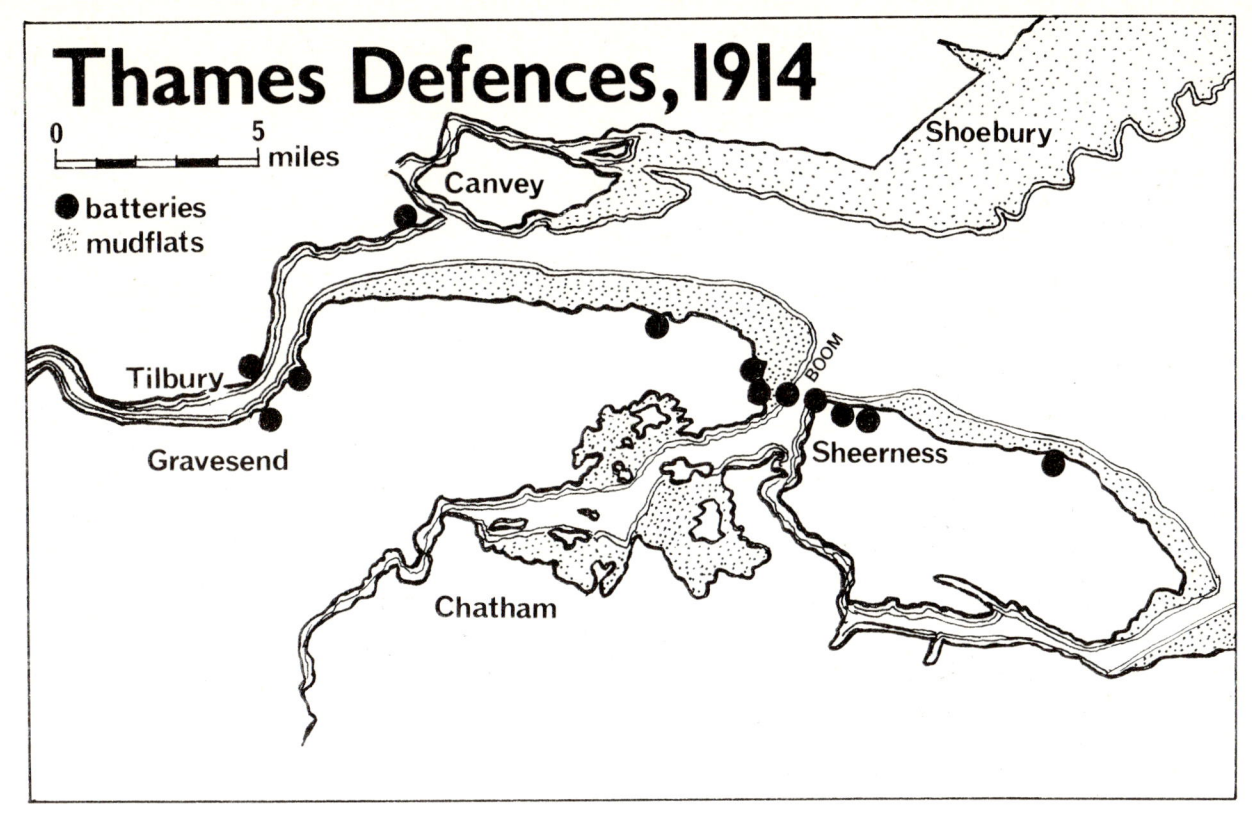

The First World War

At the time when the Thames defences had been in the process of conversion to breech-loading, the traditional fear of France was gradually supplemented and finally replaced by concern at German imperial ambitions and naval expansion. The popular press as well as the political and military elite increasingly saw Germany rather than France as a possible future enemy. The Entente Cordiale in 1904 removed France as a potential adversary.

When war with Germany came in 1914, the Thames forts were manned in accordance with previously worked out mobilisation plans. Minefields were established and the river was made subject to the special restrictions on ship movements which operated in war. These restrictions were enforced by the River Examination Service, using HMS Champion, an old wooden warship moored in the middle of the river between Coalhouse and Cliffe forts. Several requisitioned Sun tugs were based on her and routinely met incoming vessels to check papers and to receive the daily password. As an added security measure, Coalhouse Fort was designated an Examination Battery and was empowered to fire a warning shot across the bows of any suspect vessel or any which refused to stop.

9.2 inch gun at Fletcher Battery, Isle of Sheppey, during the First World War   (RE Corps Library)

Field gun emplaced as part of the land defence line in advance of Chatham during the First World War   *(R.E. Corps Library)*

So far as the forts themselves were concerned, it seems certain that all remaining RML guns had been removed before 1914. As to the BL armaments, the situation is ambiguous. Some official records have been interpreted to suggest that only Coalhouse Fort of the forts in the Gravesend Reach-Lower Hope Point section of river was actually armed during the war. However, guns were certainly retained at New Tavern where there was a Royal Artillery garrison. Tilbury Fort too had a garrison and it has been suggested that some guns remained. Two of the 6-in. guns at Coalhouse Fort were transferred to Cliffe Fort in 1914, where new roof emplacements were built to receive them. All the QF guns at Coalhouse had previously been withdrawn and their sites obscured by electric light emplacements. Whether the small QF battery to the SW of Coalhouse was armed during the war is uncertain but it had by then also been partially obscured by electric light emplacements, including one having two beam positions. East Tilbury Battery appears to have been disarmed.

The QF battery at Shornemead was retained and rearmed with 12-pdr. weapons but the one at Lower Hope seems to have been abandoned. Finally, the two 9.2-in. weapons at Slough Fort were transferred to Fletcher Battery on the Isle of Sheppey in 1918, another sign of the continuing trend towards siting the heavy armament for defence of the Thames further and further downstream. The Fire Command Post for the Thames defences was sited at Grain.

Shore gun defence was just part of the arrangements for the protection of the river — several interdiction flotillas of destroyers, torpedo boats and submarines were based at Chatham and Sheerness to attack enemy raiders before they came within range of the batteries. However, the advances in military aviation meant that the defenders had to meet a threat from another dimension. Anti-aircraft batteries were set up in the lower Thames area both for the purposes of the local defence of vulnerable and important points such as oil stores and jetties and as part of the much wider strategy for the air defence of the approaches to the capital, within the London Air Defence Area. In 1916, guns on the parade of Tilbury Fort managed to damage a German Zeppelin. This was the Reichsmarine L15. Leaking gas, it fell into the sea a mile from the Kentish Knock lightship. Zeppelins had bombed Gravesend in 1915. During raids, sometimes the small gondolas of the Zeppelins could be seen from the ground if they had been lowered below the clouds on the end of a long cable to give navigational directions to the mother ship. Bombers also used the line of the Thames as a navigational aid to get them to London. Anti-aircraft guns were a secondary inner line of air defence. The first line comprised squadrons of fighter interceptors based at various airfields such as Dartford, Grain, Woodchurch and Southend. Not only were the fighters intended to combat other aeroplanes and airships but in the event of an invasion were earmarked for attacking any naval forces making an attempt on the Thames as well as for straffing a landing force.

As to precautions against an actual landing attempt, the coast defence guns would have been used where they could be brought to bear but the defence against an actual disembarkation would have devolved on the Thames Local Force. This comprised infantry, artillery and cavalry and was deployed in various places on both sides of the river near the estuary. It was expected to operate in accordance with the Thames and Medway Defence Plan which had been worked out before the war. To strengthen the defence, there were various entrenched positions and strong-points. At Shoeburyness for example, there was an extensive system of trenches, barbed wire entanglements, blockhouses and pillboxes. There were other such defences on the south side of the river in the Hoo Peninsula, including a small complex at Grain, adjacent to the batteries there. On the Isle of Sheppey, there was a defensive line covering the land approaches to the Sheerness naval base. There was also a long line between the Swale and the ground to the north of Maidstone. The general strategy was for the raiders or invaders to be held long enough by the local forces to buy time for a massive and decisive counter-blow by the Main Central Force, held in reserve. The German general staff had framed plans for a landing in the Thames as part of a general invasion of England and documentary evidence makes clear that they regarded its defences as formidable. However, no attack ever came.

One point of interest during the war was the military communication between Gravesend and Tilbury. This consisted of a bridge of boats with a movable middle section and was sited in the same place as the boom defence of 1588.

During a post-war rationalisation, it appears that only Coalhouse of the Thames forts continued to be armed although one gun was retained at New Tavern Fort for a few years for the local Territorial Army unit to practice their gun drill. The entire reliance for heavy counter-bombardment defence had already passed to the long-range guns at Grain and on the Isle of Sheppey.

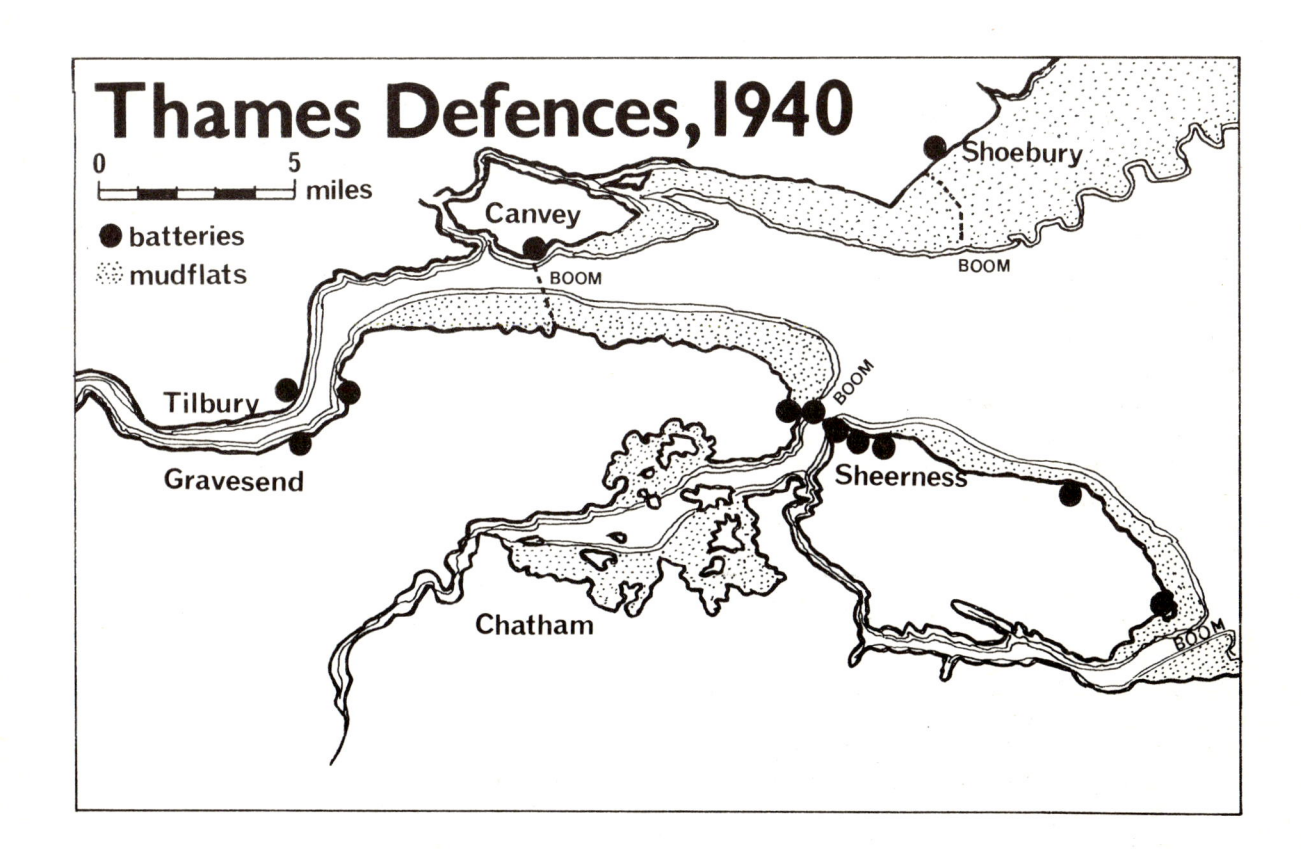

The Second World War

During the Second World War, the inter-war reliance on long-range batteries located downstream was consolidated, Canvey Battery for 2 x 6-in. guns having been built on the north shore at the time of the Munich Crisis in 1938. A third 9.2-in. gun was added at Fletcher Battery on Sheppey in 1941. There were, however, two important advances in fire control techniques which gave the defences a greater potency: the Fortress System which came first and radar.

Although still vital for close-defence, the old arrangement of individual battery range and position finders conceived in the 1880s had become less effective for counter bombardment work as the range of guns increased. Therefore a new arrangement known as the Fortress System was introduced. Under this system, developed in the 1930s, the range of vision and of precision was greatly extended by a series of Fortress Observation Posts along the coastline to cover the water within range of the guns. These transmitted bearings and ranges gained from observation to a central Fortress Plotting Room where the vessel was tracked on a chart known as a Fortress Plotter. From here the co-ordinates were telephoned or telegraphed to the individual batteries which then possessed all the information needed to engage the enemy, even though the target might be so far away as to be invisible to the Battery Commander. Another was the replacement of the old Watkin Dials for communicating ranges etc for guns by the more effective and reliable Magslip system. The Thames and Medway was divided into three Fire Commands: Counter Bombardment; Thames; Medway.

The development of radar was a very significant further advance since it was without the defects of optical observation which had to rely on good visibility but employed radio waves to find the target and to establish its range and position. It also had greater range than was possible with the telescope of a Depression Position Finder. The first Sets with fixed antennae were introduced to the Thames in 1941 but these were soon succeeded by more effective devices having movable aerials. Radar was integrated into the Fortress System and the same plotting room would be used for visually discovered bearings, radar plots or for both.

The long-range counter bombardment guns required supplementing by other 'close defence' measures to counter an actual landing attempt. This was effected by the construction of roofed over emergency batteries near the riverline, pillboxes both on the riverline and inland, entrenchments, barbed wire entanglements, minefields and anti-tank blocks. Emergency batteries at Coalhouse Fort and Shornemead (2 x 5.5-in. at each) and at Shellness and Shoeburyness (2 x 6-in. at each) were disguised with huge camouflage nets draped over them. They were each self-defensible, being surrounded by a defensive perimeter, in the case of Coalhouse, the fort itself. As to the pillboxes, many of which can still be seen, any impression of uniformity of design is illusory. They display variety both in design and in the use of materials (brick, concrete and steel). Several different weapons could be deployed in them according to type — rifles, machine-guns and even light anti-tank guns. The last two were either held by mountings which fitted into slots in the loopholes or supported on substantial concrete tables behind them. Few traces exist today of the barbed wire obstacles or entrenchments but some of the anti-tank blocks survive in various places. Southend had 1805 blocks along its seafront and two of them have been preserved and identified with a memorial plaque. Traces of others can also be seen along the front.

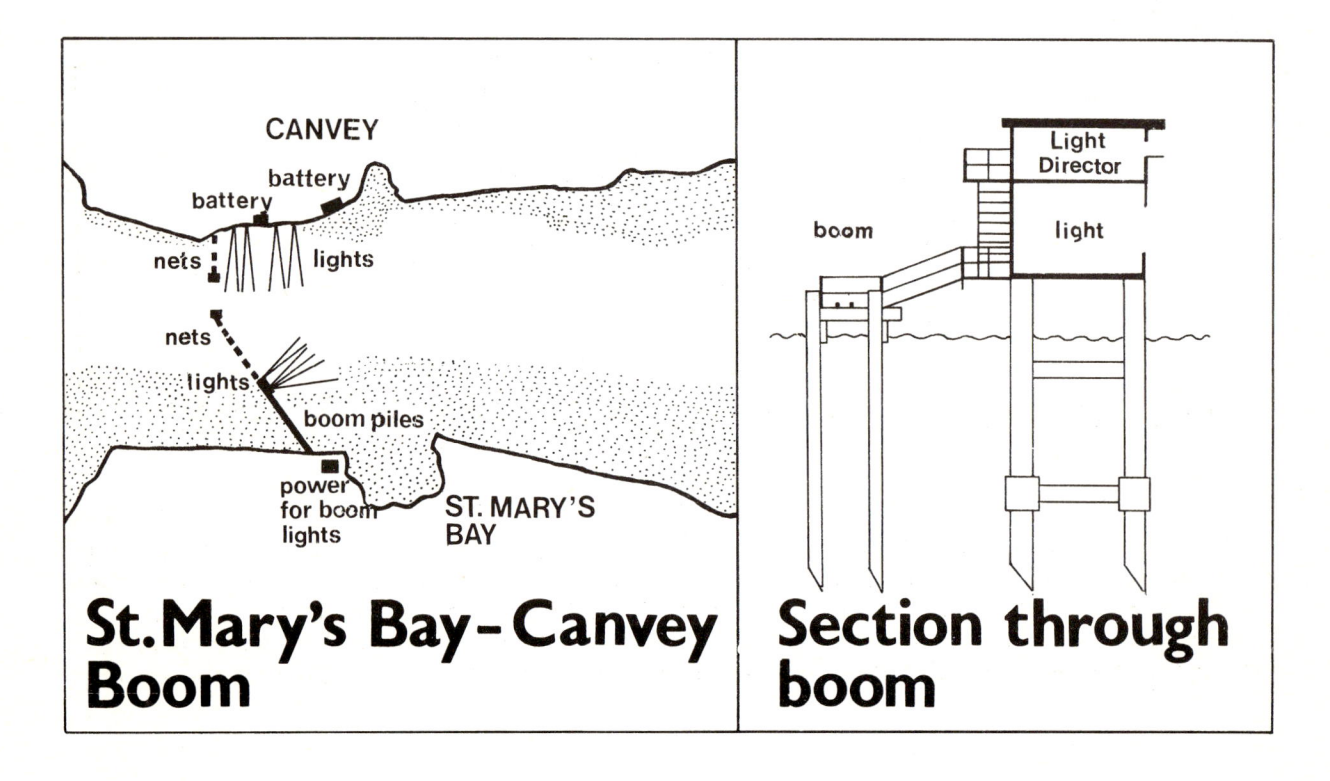

St. Mary's Bay – Canvey Boom

Section through boom

As an addition to the defences on land it was considered expedient to back them up with physical obstacles in the river itself. The first of these was a very long boom extending south from Shoeburyness. Further up was a 1¾-mile boom between St. Mary's Bay on the Kent shore and Canvey Island in Essex. It was constructed of wooden piles and had a movable middle section for shipping to pass through after challenge and inspection. Some 1,400 yards out from the Kentish shore and reached by a Decauville narrow-gauge railway on top of the boom, were three Defence Electric Light emplacements and a Directing station. These were built on stilts in front of the boom and were positioned just above water level. They were powered from generators at St. Mary's Bay. The Canvey Island end of the boom was defended by a battery of 2 twin 6-pdr. quick-firers nearby. Also on the shore of Canvey Island were four searchlights aligned to cover the water in front of the boom, especially the movable middle section. Upstream were observation minefields, controlled from at least two observation towers at Coalhouse Fort (where there was also a minefield radar tower) and Holehaven and, at the latter place, a battery of six torpedo tubes on the riverline. These were actually fired across the river on one occasion as a preventive measure when a burning tug drifting downstream seemed about to collide with the jetty on which the filled tubes were mounted. Defence from the riverbank was a military responsibility but defensive measures on the river itself was a naval one.

The Thames Auxiliary naval force was charged with patrolling the river. This force consisted of 35 small river tugs and motor launches armed respectively with 6-pdr. and .303-in. Hotchkiss machine guns, operating from bases at Dagenham, Greenhithe, Tilbury, Cliffe and Holehaven. It was the responsibility of this force to report and buoy parachute mines and to attack enemy seaborne or airborne forces in liaison with heavier units. The Cliffe base was actually Cliffe Fort, which was re-activated and re-armed with 2 x 4-in. guns for close defence.

Both the Thames and Medway contained underwater ranges for testing the effectiveness of ships' demagnetisation against the German magnetic mine.

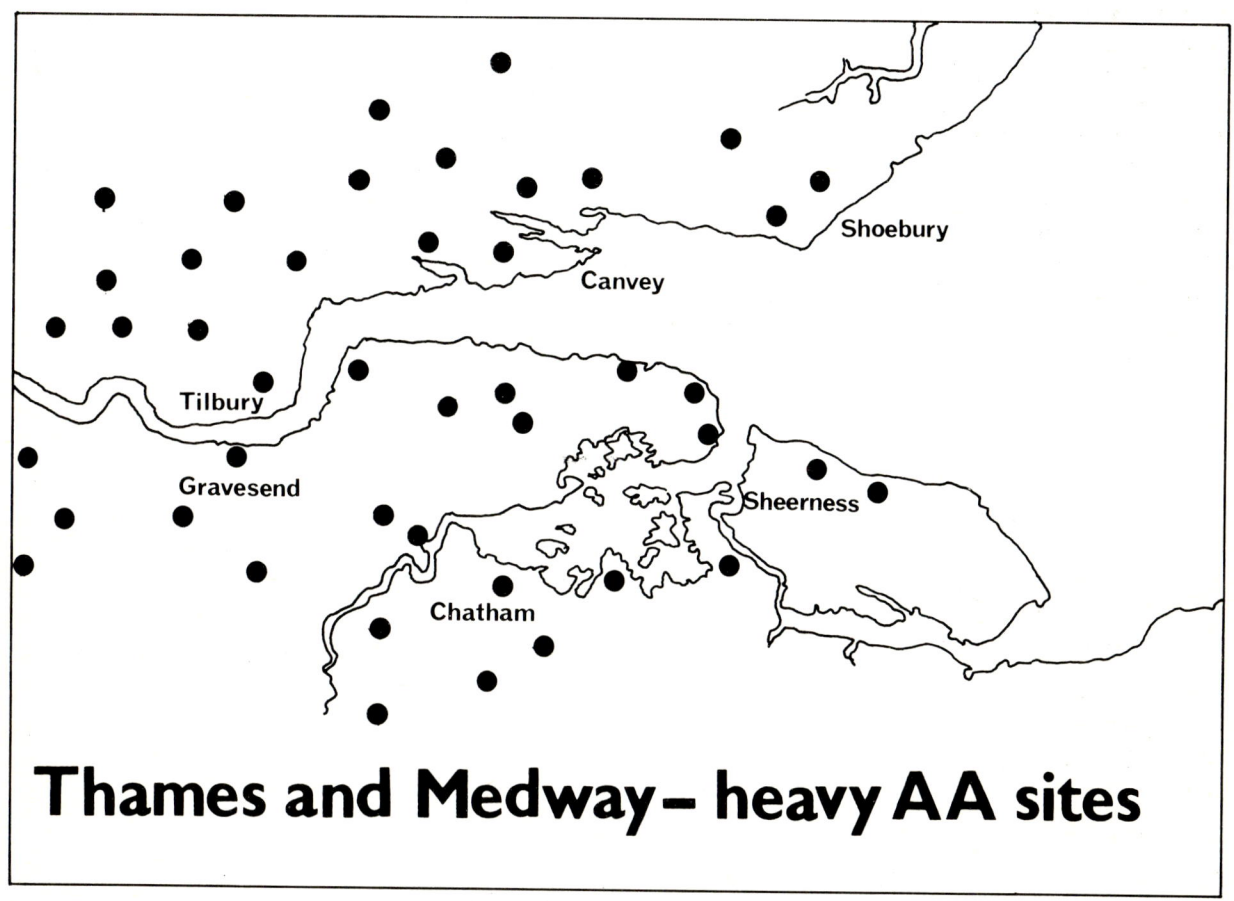

Thames and Medway – heavy AA sites

Where experience gained during the First World War had merely hinted at the possibilities of airpower, the Second World War showed it to be a most potent force. Air attacks on Britain were on a scale more vast and sustained than in the First World War and in addition to the formation of bases for fighter interceptors, this led to an extensive system of anti-aircraft artillery batteries, searchlights and radar installations many of which were established in the Thames and Medway hinterland. Early on, the German airforce exploited the gaps in the AA gun defence offered by the Thames estuary to use it as an axis of approach to London. At first, various 'flak' ships patrolled the estuary but a more permanent provision was called for. This took the form of the offshore forts, designed by Maunsell and positioned during 1942-3, some miles out into the estuary. There were two types of fort: the 'Naval' Fort (4), manned by Royal Navy and Royal Marine personnel, which consisted of two hollow concrete drums (containing accommodation and stores) bridged by a metal platform containing the guns, fire control and radar; the 'Army' forts (3), which comprised seven concrete and steel towers joined by metal catwalks. Five of the towers mounted guns on their roof, a sixth was the command post and the seventh mounted a searchlight. Both types of fort were actually built on land — some of them at Gravesend — and towed out into the estuary and sunk on the riverbed. While their 34 guns were a most welcome addition to AA gun defence, the extra radar cover which was provided by the naval forts was probably of equal value.

Air defence was not the only role of the offshore forts however. They may legitimately be regarded as 'outworks' of the coastal defences because one of their defined purposes was to prevent enemy E-Boats from carrying out raids on shipping and coastal targets in the estuary.

If the enemy had succeeded in penetrating the shore-based and riverline defences he would then have encountered the hinterland pillboxes and entrenchments, road blocks etc deployed inland to hinder and if possible arrest an advance into the hinterland. All these static defences were an adjunct to the assigned field forces which consisted of both regular and Home Guard units. The German invasion plan 'Operation Sea Lion' called for a landing in East Kent and East Sussex and a rapid advance west to achieve within a few days a front whose right flank pivoted on the south bank of the Thames at Gravesend and whose left rested on the south coast near Southampton. An assault landing on the Thames estuary, either by paratroops, marines or both was required to secure the position but there is little evidence that large scale landings on the coast north of the Thames were envisaged at this stage. Fortunately, Hitler had failed in the Battle of Britain to win the air superiority needed before an invasion could be launched and he soon gave up his ideas of invasion and turned his army against Russia instead. Provision continued to be made against invasion however. During the Second World War, the Thames saw no lack of action against air raids, minelaying (one oil tanker was destroyed by an air dropped mine in 1942) and E-Boats did on occasion succeed in penetrating into the estuary but without inflicting any serious damage.

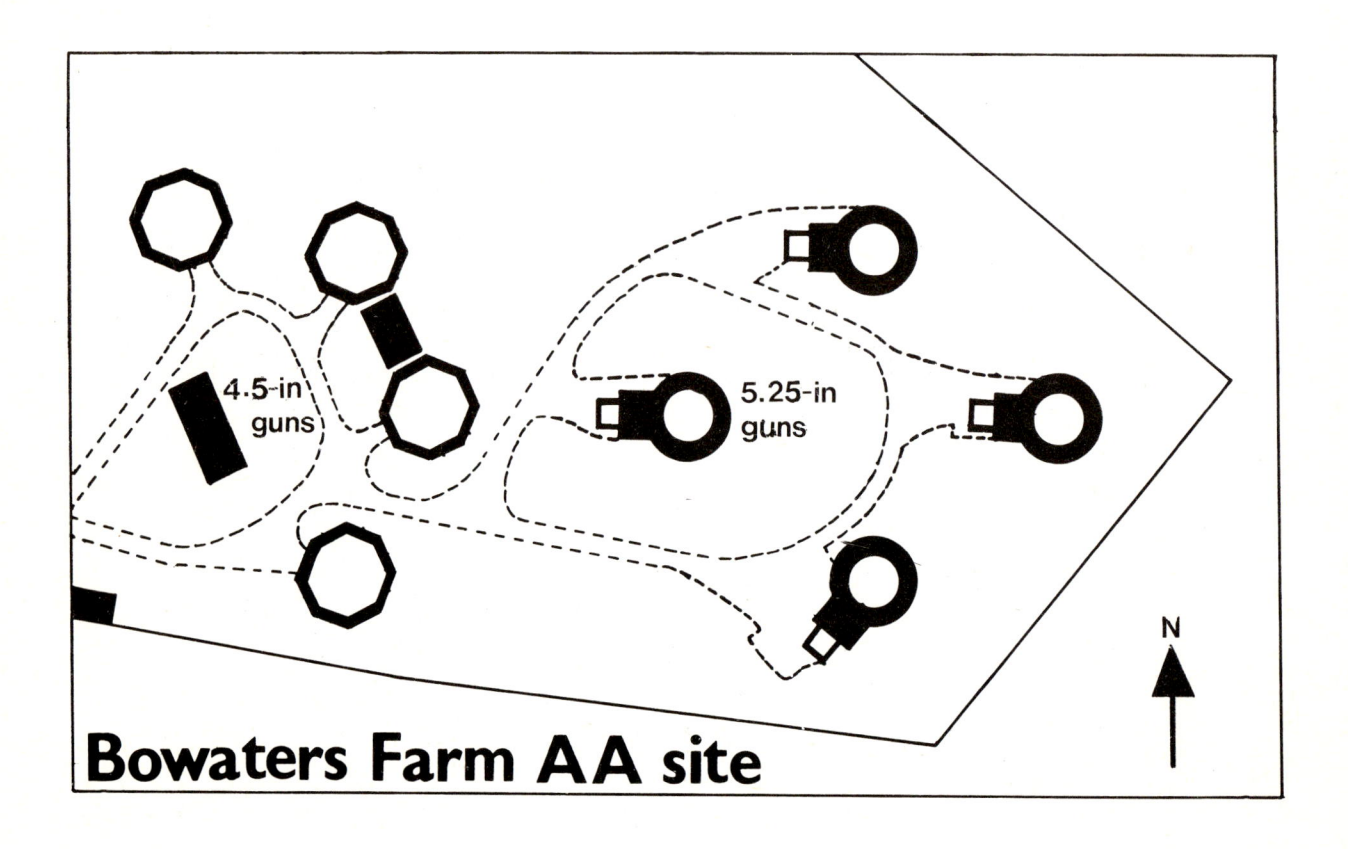

Bowaters Farm AA site

The Final Years

After the end of the Second World War, the Thames and Medway defences, as coast defences everywhere in the United Kingdom, were placed into a state of care and maintenance, although the Territorial Army gunners of 516 Coast Regiment (from 1947, 415 Coast Regiment (Thames and Medway)) engaged in strenuous training programmes on the guns which remained in position. As it happened, the days of the defences were numbered, for on 17th February 1956, the Minister of Defence announced in Parliament that the coast artillery branch of the British Army would be brought to an end. It was then argued that the development of more effective airpower and of guided missiles had finally rendered coast artillery obsolete. It was considered that attack from the sea could be more reliably defended against by the Royal Navy and Royal Air Force and, if necessary, by mobile guns deployed when and where needed. One of the last acts in the Thames was, oddly enough, an inspection of the lift equipment at Canvey Battery in the month following the Minister's announcement — a fact recorded in painted lettering at the lower ends of the lifts. However, from 1st January 1957 British coast artillery was abolished. After disarming in 1957-8, the Thames was left with its forts and battery buildings gradually decaying and becoming mute reminders in brick, stone, concrete and iron, of a branch in the Army and a system of defence which had existed for over 400 years. Fortunately, several of these forts — Tilbury, Coalhouse, New Tavern and Gravesend blockhouse — were subsequently restored and with the schemes of interpretation currently in progress, they are starting to live again.

Restoration in progress — remounting of smooth-bore cannon at New Tavern Fort (Mrs Y. Parker)

MAIN EXTANT SITES

North bank

Tilbury Fort, 1684 (TQ 552.755)

Pentagonal, bastioned fort with double water-filled ditch and outworks. Later (Victorian and early twentieth century) additions. English Heritage Monument: open standard hours.

East Tilbury QF Battery, 1893 and 1897 (TQ 691.765)

Four concrete gun emplacements, magazines and later electric light emplacements on mound behind river wall. Open site on edge of Thurrock Borough Council recreation area.

Radar tower, 1941 (TQ 689.763)

Hexagonal tower on riverline. Open site on edge of Thurrock Borough Council recreation area.

Coalhouse Fort, 1874 (TQ 691.768)

Curved front of casemates and open battery with underlying magazines and indented defensible barracks closing the rear. Later 19th and early 20th century additions on roof. Detached minefield control tower in front of fort. Restoration in progress. Visiting on public holidays and on some weekends. Minefield tower on open site.

East Tilbury Battery, 1891-2 (TQ 687.774)

Elongated wing battery of Coalhouse Fort, consisting of 6 concrete gun emplacements, magazines, surrounded by a ditch and fence. Overgrown and derelict. On private property, fenced off.

Bowaters Farm AA Battery, 1940 (TQ 678.771)

Eight concrete gun emplacements, command post, magazines and barracks. Private property. Permission to visit must be sought.

Hadleigh Castle, 1360s (TQ 810.861)

Ruin. English Heritage monument, open standard hours.

Shoeburyness range, 1855 (TQ 953.857)

Magazines, several emplacements, electric light emplacements and part of boom. Ministry of Defence property. No public access.

South bank

Gravesend Blockhouse, 1540 (TQ 650.744)

Remains of D-shaped brick blockhouse. Can be seen in the riverside gardens of the Clarendon Royal Hotel. The hotel itself was the Duke of York's HQ in 1665 but has been much altered since.

New Tavern Fort, 1783 (TQ 653.743)

18th century earthen battery, remodelled in 1868-72 and in 1904. Emplacements and magazines from these remodellings survive. Open on public holidays and on Sunday mornings.

Shornemead Fort, c 1870 (TQ 693.748)

Similar to Coalhouse Fort but smaller. Casemated front survives. Ministry of Defence property but on open site.

Cliffe Fort, c 1870 (TQ 707.767)

Similar to Coalhouse and Shornemead forts, with, with QF emplacements (1895) and 6-inch positions (WW1) on roof. The fort is owned by Blue Circle Industries Ltd from which permission for visiting must be gained. Traces of the Brennan Torpedo Station (1888) in front of the fort are outside the fenced off area.

Cooling Castle, 1380s (TQ 753.750)

Ruin. On private property.

Inner Thames Boom, 1940 (TQ 788.791)

Engine room, dining hall and cookhouse remain. Open site behind river wall.

Slough Fort, 1867 (TQ 835.788)

Small semi-circular casemated work with wing batteries of 1905. Within perimeter of riding school.

Offshore Forts, 1942

Nore 51°25.45N 0°50.00E
Red Sands 51°28.62N 0°59.60E   Army
Shivering Sands 51°29.95N 1°04.48E

Knock John 51°33.7N 1°09.82E
Roughs 51°33.66N 1°28.93E
Sunk Head 51°46.51N 1°30.21E   Navy
Tongue Sands 51°29.55N 1°22.11E

Attempts to gain entry are not advised. Hazardous sites.

Offshore Fort (Army) during the Second World War (*Skyfotos*)

MEDWAY DEFENCES

The few traces of Grain Fort, Grain Battery and Wing Battery which survive are easy of access, being on reasonably open sites. Grain Tower, with its later additions, remains intact and may be reached along a hardway from the shore. For more details of the Grain defences see P. MacDougall, 'The Isle of Grain Defences', Kent Defence Research Group (1980).

On the Isle of Sheppey, Garrison Point Fort and other batteries are extant but much of this is within the security area of the Medway Ports Authority from whom permission to visit must be obtained. To the east of Garrison Point are some fragments of Ravelin and Barton's Point Batteries. Much further to the east is Fletcher Battery, within the area of a holiday camp. To the east of that are some remains of military structures at Warden Point and, not far away, a concrete 'early warning' acoustic mirror which has slid down onto the beach from the cliffs above. For more details of the Medway and Sheerness defences the reader is referred to K.R. Gulvin, 'The Medway Forts', Medway Military Research Group (1977).

Other information about the Medway forts which fired on the Thames may be found in Ian V. Hogg, 'Coast Defences of England and Wales, 1856-1956', Newton Abbot (1974).

SOURCES

Many sources were used in the research for this book. The main ones were as follows:

Primary

Calendars of State Papers (Domestic);
Documents and maps at the Public Record Office in the WO 32, 33, 44, 55, 78 and 199 series as well as Cabinet Papers of the later nineteenth and early twentieth centuries;
Letter books at the RE Corps Library, Chatham;
Diaries of John Evelyn and Samuel Pepys;
'Report of the Commissioners appointed to consider the Defences of the United Kingdom', London, 1860 and progress reports of 1867, 1868 and 1869;
Army Estimates in Parliamentary Accounts;
Maps and plans in the collections of Gravesham Borough Library, the Kent Archives Office and the British Library;
Various official reports and returns of the later nineteenth and early twentieth centuries in the Old War Office Library, London and Naval Historical Branch, London;
Photographic collection of the Imperial War Museum;
Journals and professional papers of the Royal Artillery, Royal Engineers and Journal of the Royal United Services Institute.

Secondary

R.P. Cruden, 'History of Gravesend', London, 1843;
A.D. Saunders, 'Tilbury Fort and the Development of Artillery Fortification in the Thames Estuary', Antiquaries Journal, XL 1960;
J.D. Wilson, 'Later Nineteenth Century Defences of the Thames', Journal of the Society for Army Historical Research, XII, 1963;
Victor Smith, 'The Artillery Defences at Gravesend', Archaeologia Cantiana, LXXXIX, 1974;
Ian V. Hogg, 'Coast Defences of England and Wales 1856-1956', Newton Abbot, 1977;
Various issues of 'Panorama', the journal of Thurrock Local History Society.